METHODS IN MOLECULAR BIOLOGY

For further volumes:
http://www.springer.com/series/7651

For over 35 years, biological scientists have come to rely on the research protocols and methodologies in the critically acclaimed *Methods in Molecular Biology* series. The series was the first to introduce the step-by-step protocols approach that has become the standard in all biomedical protocol publishing. Each protocol is provided in readily-reproducible step-by step fashion, opening with an introductory overview, a list of the materials and reagents needed to complete the experiment, and followed by a detailed procedure that is supported with a helpful notes section offering tips and tricks of the trade as well as troubleshooting advice. These hallmark features were introduced by series editor Dr. John Walker and constitute the key ingredient in each and every volume of the *Methods in Molecular Biology* series. Tested and trusted, comprehensive and reliable, all protocols from the series are indexed in PubMed.

Ferroptosis

Methods and Protocols

Edited by

Guido Kroemer

Centre de Recherche des Cordeliers, Université de Paris, Paris, France

Daolin Tang

Department of Surgery, University of Texas Southwestern Medical Center, Dallas, TX, USA

Humana Press

Editors
Guido Kroemer
Centre de Recherche des Cordeliers
Université de Paris
Paris, France

Daolin Tang
Department of Surgery
University of Texas Southwestern Medical Center
Dallas, TX, USA

ISSN 1064-3745 ISSN 1940-6029 (electronic)
Methods in Molecular Biology
ISBN 978-1-0716-3432-5 ISBN 978-1-0716-3433-2 (eBook)
https://doi.org/10.1007/978-1-0716-3433-2

This Humana imprint is published by the registered company Springer Science+Business Media, LLC, part of Springer Nature.
The registered company address is: 1 New York Plaza, New York, NY 10004, U.S.A.

Preface

Ferroptosis is a type of regulated cell death that is driven by iron-dependent lipid peroxidation. Over the past decade, ferroptosis has been discovered to play a critical role in the pathogenesis of various diseases such as neurodegeneration, cancer, and ischemia-reperfusion injury. As the importance of ferroptosis for human health and disease becomes increasingly evident, the demand for reliable and sensitive methods for detecting and measuring ferroptosis has rapidly grown.

This book, *Ferroptosis: Methods and Protocols*, provides a comprehensive collection of experimental protocols for investigating ferroptosis in different systems, including cultured cells, animal models, and human tissues. Written by leading researchers in the field, each chapter in this book describes the methodology used to study various aspects of ferroptosis, from the detection of lipid peroxidation to the measurement of glutathione peroxidase activity and the evaluation of mitochondrial morphology.

This book is a valuable resource for researchers who are interested in studying ferroptosis in different contexts, including basic research, drug discovery, and clinical applications. It covers a wide range of experimental techniques, from basic molecular biology methods such as quantitative PCR and immunoblotting to advanced imaging techniques such as transmission electron microscopy and confocal fluorescence microscopy. The protocols presented in this book have been developed, tested, and optimized by experienced researchers in the field. Each of them is accompanied by detailed step-by-step instructions, troubleshooting tips, and critical notes.

We are confident that this book will serve as a practical guide for researchers who explore the fascinating field of ferroptosis, thus contributing to the understanding and treatment of major human diseases.

Dallas, TX, USA *Daolin Tang*
Paris, France *Guido Kroemer*

Contents

Contributors

SRIDHAR BAMMIDI • *Department of Ophthalmology, University of Pittsburgh School of Medicine, Pittsburgh, PA, USA*

FANGQUAN CHEN • *DAMP Laboratory, The Third Affiliated Hospital of Guangzhou Medical University, Guangzhou, Guangdong, China*

RUOCHAN CHEN • *Department of Infectious Diseases, Xiangya Hospital, Central South University, Changsha, Hunan, China; Hunan Key Laboratory of Viral Hepatitis, Xiangya Hospital, Central South University, Changsha, Hunan, China*

ZHIYANG CHEN • *Key Laboratory of Regenerative Medicine of Ministry of Education, Institute of Aging and Regenerative Medicine, Jinan University, Guangzhou, China*

ZHANGSHUAI DAI • *Key Laboratory for Regenerative Medicine, Ministry of Education, College of Life Science and Technology, Jinan University, Guangzhou, China*

CHAO DENG • *Department of Oncology, The Second Xiangya Hospital, Central South University, Changsha, Hunan, China*

SAYAN GHOSH • *Department of Ophthalmology, University of Pittsburgh School of Medicine, Pittsburgh, PA, USA; Wilmer Eye Institute, The Johns Hopkins University School of Medicine, Baltimore, MD, USA*

JINGYI GUO • *CAS Key Laboratory of Regenerative Biology, Joint School of Life Sciences, Guangzhou Institutes of Biomedicine and Health, Chinese Academy of Sciences, Guangzhou Medical University, Guangzhou, China; Innovation Centre for Advanced Interdisciplinary Medicine, The Fifth Affiliated Hospital of Guangzhou Medical University, Guangzhou, China*

JAMES T. HANDA • *Wilmer Eye Institute, The Johns Hopkins University School of Medicine, Baltimore, MD, USA*

RONG-RONG HE • *Guangdong Engineering Research Center of Chinese Medicine & Disease Susceptibility, Jinan University, Guangzhou, China*

STACEY HOSE • *Department of Ophthalmology, University of Pittsburgh School of Medicine, Pittsburgh, PA, USA*

QIAN HU • *Key Laboratory of Regenerative Medicine of Ministry of Education, Institute of Aging and Regenerative Medicine, Jinan University, Guangzhou, China*

JUNQI HUANG • *Key Laboratory for Regenerative Medicine, Ministry of Education, College of Life Science and Technology, Jinan University, Guangzhou, China*

SIYUAN HUANG • *Department of Trauma Medical Center, State Key Laboratory of Trauma, Burns and Combined Injury, Daping Hospital, Army Medical University, Chongqing, China*

GA-RAM HWANG • *Department of Pathology, Renaissance School of Medicine, Stony Brook University, Stony Brook, NY, USA*

LILI JIANG • *Affiliated Cancer Hospital & Institute of Guangzhou Medical University, Guangzhou, China; Guangzhou Municipal and Guangdong Provincial Key Laboratory of Protein Modification and Degradation, School of Basic Medical Science, Guangzhou Medical University, Guangzhou, China*

JINGFANG JU • *Department of Pathology, Renaissance School of Medicine, Stony Brook University, Stony Brook, NY, USA; The Northport Veteran's Administration Medical Center, Northport, NY, USA*

ZHENYU JU • *Key Laboratory of Regenerative Medicine of Ministry of Education, Institute of Aging and Regenerative Medicine, Jinan University, Guangzhou, China*
RUI KANG • *Department of Surgery, UT Southwestern Medical Center, Dallas, TX, USA*
JINGBO LI • *The Third Xiangya Hospital, Central South University, Changsha, China*
QIAN LI • *Department of Biochemistry and Molecular Biology, Capital Medical University, Beijing, China*
WENXIN LI • *CAS Key Laboratory of Regenerative Biology, Joint School of Life Sciences, Guangzhou Institutes of Biomedicine and Health, Chinese Academy of Sciences, Guangzhou Medical University, Guangzhou, China; Guangdong Provincial Key Laboratory of Stem Cell and Regenerative Medicine, China-New Zealand Joint Laboratory on Biomedicine and Health, CUHK-GIBH Joint Research Laboratory on Stem Cells and Regenerative Medicine, Institute for Stem Cell and Regeneration, Guangzhou Institutes of Biomedicine and Health, Chinese Academy of Sciences, Guangzhou, China; University of Chinese Academy of Sciences, Beijing, China*
JIAO LIU • *DAMP Laboratory, The Third Affiliated Hospital of Guangzhou Medical University, Guangzhou, Guangdong, China*
XINGGUO LIU • *CAS Key Laboratory of Regenerative Biology, Joint School of Life Sciences, Guangzhou Institutes of Biomedicine and Health, Chinese Academy of Sciences, Guangzhou Medical University, Guangzhou, China; Guangdong Provincial Key Laboratory of Stem Cell and Regenerative Medicine, China-New Zealand Joint Laboratory on Biomedicine and Health, CUHK-GIBH Joint Research Laboratory on Stem Cells and Regenerative Medicine, Institute for Stem Cell and Regeneration, Guangzhou Institutes of Biomedicine and Health, Chinese Academy of Sciences, Guangzhou, China; Centre for Regenerative Medicine and Health, Hong Kong Institute of Science & Innovation, Chinese Academy of Sciences, Hong Kong, SAR, China*
LIFANG MA • *Shanghai Institute of Thoracic Oncology, Shanghai Chest Hospital, Shanghai Jiao Tong University School of Medicine, Shanghai, China; Department of Clinical Laboratory, Shanghai Chest Hospital, Shanghai Jiao Tong University School of Medicine, Shanghai, China*
NATHCHAR NAOWAROJNA • *Westlake Four-Dimensional Dynamic Metabolomics (Meta4D) Laboratory, Westlake Laboratory of Life Sciences and Biomedicine, HangzhouZhejiang, China; School of Life Sciences, Westlake University, HangzhouZhejiang, China; Research Center for Industries of the Future, Westlake University, HangzhouZhejiang, China; Westlake Institute for Advanced Study, Hangzhou, Zhejiang, China*
DANMIN SHEN • *Department of Biochemistry and Molecular Biology, Capital Medical University, Beijing, China; Department of Neurology, Beijing Children's Hospital, Capital Medical University, National Center for Children's Health, Beijing, China*
DEBASISH SINHA • *Department of Ophthalmology, University of Pittsburgh School of Medicine, Pittsburgh, PA, USA; Wilmer Eye Institute, The Johns Hopkins University School of Medicine, Baltimore, MD, USA*
YUJIE SU • *CAS Key Laboratory of Regenerative Biology, Joint School of Life Sciences, Guangzhou Institutes of Biomedicine and Health, Chinese Academy of Sciences, Guangzhou Medical University, Guangzhou, China; Guangdong Provincial Key Laboratory of Stem Cell and Regenerative Medicine, China-New Zealand Joint Laboratory on Biomedicine and Health, CUHK-GIBH Joint Research Laboratory on Stem Cells and Regenerative Medicine, Institute for Stem Cell and Regeneration, Guangzhou Institutes of Biomedicine and Health, Chinese Academy of Sciences, Guangzhou, China; University of Chinese Academy of Sciences, Beijing, China*

WAN-YANG SUN • *Guangdong Engineering Research Center of Chinese Medicine & Disease Susceptibility, Jinan University, Guangzhou, China*

DAOLIN TANG • *Department of Surgery, UT Southwestern Medical Center, Dallas, TX, USA*

TANIA TAO • *NHC Key Laboratory of Carcinogenesis, Cancer Research Institute and School of Basic Medicine, Central South University, Changsha, Hunan, China*

YONGGUANG TAO • *NHC Key Laboratory of Carcinogenesis, Cancer Research Institute and School of Basic Medicine, Central South University, Changsha, Hunan, China; Department of Pathology, Xiangya Hospital, Central South University, Changsha, Hunan, China*

FENGXIANG WANG • *Westlake Four-Dimensional Dynamic Metabolomics (Meta4D) Laboratory, Westlake Laboratory of Life Sciences and Biomedicine, HangzhouZhejiang, China; School of Life Sciences, Westlake University, HangzhouZhejiang, China; Research Center for Industries of the Future, Westlake University, HangzhouZhejiang, China*

JIAYI WANG • *Shanghai Institute of Thoracic Oncology, Shanghai Chest Hospital, Shanghai Jiao Tong University School of Medicine, Shanghai, China; Department of Clinical Laboratory, Shanghai Chest Hospital, Shanghai Jiao Tong University School of Medicine, Shanghai, China; The Third Xiangya Hospital, Central South University, Changsha, China*

MANYUAN WANG • *College of Health Science and Technology, Shanghai Jiao Tong University School of Medicine, Shanghai, China*

MENGFEI WANG • *CAS Key Laboratory of Regenerative Biology, Joint School of Life Sciences, Guangzhou Institutes of Biomedicine and Health, Chinese Academy of Sciences, Guangzhou Medical University, Guangzhou, China; Guangdong Provincial Key Laboratory of Stem Cell and Regenerative Medicine, China-New Zealand Joint Laboratory on Biomedicine and Health, CUHK-GIBH Joint Research Laboratory on Stem Cells and Regenerative Medicine, Institute for Stem Cell and Regeneration, Guangzhou Institutes of Biomedicine and Health, Chinese Academy of Sciences, Guangzhou, China; University of Chinese Academy of Sciences, Beijing, China*

RONG WANG • *Guangdong Engineering Research Center of Chinese Medicine & Disease Susceptibility, Jinan University, Guangzhou, China*

YANG WANG • *Institute of Cancer Stem Cell, Dalian Medical University, Dalian, China*

ZULI WANG • *Center for Tissue Engineering and Stem Cell Research, Guizhou Medical University, Guiyang, Guizhou, China; NHC Key Laboratory of Carcinogenesis, Cancer Research Institute and School of Basic Medicine, Central South University, Changsha, Hunan, China; Department of Pathology, Xiangya Hospital, Central South University, Changsha, Hunan, China*

RUNLIU WU • *Department of Surgery, The Third Xiangya Hospital, Central South University, Changsha, Hunan, China*

YANGCHUN XIE • *Department of Oncology, The Second Xiangya Hospital, Central South University, Changsha, Hunan, China*

XIANGFEI XUE • *Shanghai Institute of Thoracic Oncology, Shanghai Chest Hospital, Shanghai Jiao Tong University School of Medicine, Shanghai, China*

FEI YANG • *Department of Neurobiology, Capital Medical University, Beijing, China*

MINGHUA YANG • *Department of Pediatrics, The Third Xiangya Hospital Central South University, Changsha, Hunan, China*

CHUNHUA YU • *Department of Surgery, UT Southwestern Medical Center, Dallas, TX, USA*

JOHN G. YUEN • *Department of Pathology, Renaissance School of Medicine, Stony Brook University, Stony Brook, NY, USA*

LING ZENG • *Department of Trauma Medical Center, State Key Laboratory of Trauma, Burns and Combined Injury, Daping Hospital, Army Medical University, Chongqing, China*

WANTING ZHANG • *Key Laboratory for Regenerative Medicine, Ministry of Education, College of Life Science and Technology, Jinan University, Guangzhou, China*

WENJING ZHANG • *Institute of Cancer Stem Cell, Dalian Medical University, Dalian, China*

XIAO ZHANG • *Shanghai Institute of Thoracic Oncology, Shanghai Chest Hospital, Shanghai Jiao Tong University School of Medicine, Shanghai, China*

YIFAN ZHANG • *Key Laboratory of Regenerative Medicine of Ministry of Education, Institute of Aging and Regenerative Medicine, Jinan University, Guangzhou, China*

ZHUOJUN ZHANG • *Affiliated Cancer Hospital & Institute of Guangzhou Medical University, Guangzhou, China; Guangzhou Municipal and Guangdong Provincial Key Laboratory of Protein Modification and Degradation, School of Basic Medical Science, Guangzhou Medical University, Guangzhou, China*

ZHI LIN • *Department of Pediatrics, The Third Xiangya Hospital Central South University, Changsha, Hunan, China*

XIAO ZHONG • *Department of Infectious Diseases, Xiangya Hospital, Central South University, Changsha, Hunan, China; Hunan Key Laboratory of Viral Hepatitis, Xiangya Hospital, Central South University, Changsha, Hunan, China*

LIQUN ZHOU • *Key Laboratory for Regenerative Medicine, Ministry of Education, College of Life Science and Technology, Jinan University, Guangzhou, China*

YILONG ZOU • *Westlake Four-Dimensional Dynamic Metabolomics (Meta4D) Laboratory, Westlake Laboratory of Life Sciences and Biomedicine, HangzhouZhejiang, China; School of Life Sciences, Westlake University, HangzhouZhejiang, China; Research Center for Industries of the Future, Westlake University, HangzhouZhejiang, China; Westlake Institute for Advanced Study, Hangzhou, Zhejiang, China*

Chapter 1

Fluorogenic Probes for Intracellular Iron Detection

Runliu Wu, Daolin Tang, and Rui Kang

Abstract

Iron is a crucial element required to sustain multiple biological processes, including oxygen transport, DNA synthesis, and electron transport. In living cells, iron exists as either ferrous iron (Fe^{2+}) or ferric iron (Fe^{3+}), and its redox forms are regulated by the labile iron pool. Both iron deficiency and excess can lead to a range of pathological conditions, such as anemia, cancer, neurodegenerative disorders, and ischemia and reperfusion injury. Iron overload can cause oxidative damage and even cell death, especially via ferroptosis. Impaired ferroptosis pathways are implicated in the pathogenesis of various diseases and are becoming attractive therapeutic targets. Therefore, developing methods to analyze dynamic iron changes in cells is crucial. In this chapter, we introduce several protocols that use fluorogenic iron probes (e.g., FerroFarRed, Calcein-AM, and FRET iron probe 1) to measure intracellular iron content.

Key words Iron probes, Iron content, Cell death

1 Introduction

Iron is the most abundant transition metal in the human body, playing an essential role in many metabolic processes such as oxygen metabolism, proton transfer, and enzyme synthesis [1]. Maintaining cellular iron homeostasis requires a balance between iron uptake, transport, utilization, and storage [2]. Abnormal iron abundance and redox state can lead to serious diseases. Iron overload is implicated in cancers, neurodegenerative disorders, and osteoporosis, whereas iron deficiency is associated with anemia [3, 4]. In living cells, iron primarily exists as either ferrous iron (Fe^{2+}) or ferric iron (Fe^{3+}). Intracellular Fe^{2+} overload can mediate the production of reactive oxygen species (ROS) through the Fenton reaction, resulting in severe oxidative damage to DNA, proteins, and lipids [5]. Ultimately, these events impair crucial metabolic functions, leading to cell death or tissue injury [6, 7].

The labile iron pool (LIP) is a small fraction of iron that weakly binds to proteins in cells. The LIP serves as a crucial system for maintaining iron homeostasis by being chelatable, exchangeable,

Guido Kroemer and Daolin Tang (eds.), *Ferroptosis: Methods and Protocols*, Methods in Molecular Biology, vol. 2712, https://doi.org/10.1007/978-1-0716-3433-2_1,

and redox-active [8, 9]. Any disturbance to LIP can contribute to iron overload, leading to iron-induced oxidative damage such as ferroptosis [10]. Ferroptosis is a regulated cell death process characterized by iron-dependent lipid peroxidation [11]. Excessive heme and non-heme iron, increased LIP, or exogenous iron supplementation induce or promote ferroptosis [12, 13]. This process is regulated by certain iron-containing enzymes [14–16]. In contrast, iron chelators can inhibit ferroptotic cell death. Developing methods to monitor intracellular iron concentration is necessary for understanding the physiological and pathological properties of ferroptosis and iron-related diseases, including cancer [13, 17].

Iron probes are valuable tools for measuring the iron levels in different biological samples. Chromogenic and fluorogenic probes are both widely used for detecting iron. The chromogenic method, using an iron probe, provides a convenient way to measure both Fe^{2+} and Fe^{3+} simultaneously [18]. Fluorescent iron probes have also been extensively studied due to their high selectivity, sensitivity, and simplicity. Fluorescence detection offers a real-time optical readout, which can help to analyze the dynamic changes in iron levels under different conditions, such as during ferroptosis. In this chapter, we introduce methods for measuring iron content in biological samples using iron probes.

2 Materials

1. Pipette tips.
2. 75 cm^2 cell culture flasks, glass bottom cell culture dish, clear 96-well cell culture plates.
3. Centrifuge tubes.
4. Nitrile gloves.
5. 50 mL tube.
6. Cell scraper..
7. Cell strainer.
8. PANC1 cells.
9. HepG2 cells.
10. Phosphate-buffered saline (PBS).
11. 0.25% trypsin-EDTA.
12. Dulbecco's modified eagle medium (DMEM).
13. Fetal bovine serum (FBS).
14. Hank's balanced salt solution (HBSS).
15. Dimethyl sulfoxide (DMSO).
16. Bovine serum albumin (BSA).

17. 40 μm nylon mesh.
18. Humidified incubator.
19. Benchtop centrifuge.
20. Homogenizer.
21. Sonicator.
22. Fluorescence microscope.
23. Microplate reader.
24. Flow cytometry machine.

3 Methods

3.1 Iron Probe for Measuring Fe^{2+}

Fe^{2+} is a well-known catalyst for the production of ROS, which can lead to oxidative damage of cells [5]. Monitoring the content of Fe^{2+} is crucial for understanding its role in various physiological activities and can be used to assess the process of ferroptosis. Several fluorescent probes, including FerroFarRed, FerroOrange, and mito-FerroGreen, have been designed for detecting intracellular Fe^{2+} through fluorescent microscopy or flow cytometry analysis [19]. In particular, mito-FerroGreen is used for Fe^{2+} detection in mitochondria. These probes specifically react with Fe^{2+} rather than other metal ions in living cells, and irreversibly turn into a far-red, orange, or green fluorescent substance, providing a highly sensitive method to measure the level of Fe^{2+}.

The FerroOrange and mito-FerroGreen assays share a similar protocol to FerroFarRed (*see* **Note 1**).

1. Plate PANC1 or HepG2 cells in a glass bottom dish at an appropriate density and culture overnight at 37 °C.
2. Remove the culture medium from the dish and rinse twice gently with serum-free medium or HBSS.
3. Treat cells with a Fe^{2+}-specific chelator (e.g., 20 μM desferoxamine [DFO]), ferroptosis inducer erastin (10 μM), $FeCl_3$ (50 μM), or DMSO for 12–24 h (*see* **Notes 2** and **3**).
4. Add 50 μL of DMSO to 1 vial (50 nmol) to prepare 1 mM FerroFarRed stock solution and pipette several times to completely dissolve the solid.
5. Dilute 1 mM FerroFarRed stock solution in serum-free medium to prepare a staining solution with a final concentration of 5 μM (use serum-free solution to prevent the dye from reacting with Fe^{2+} in the serum).
6. Add the staining solution to the dish and incubate for 1 h at 37 °C.

7. After staining, proceed with either (a) observation by fluorescent microscopy or (b) analysis by flow cytometry:
   (a) For observation by fluorescent microscopy:
      1. Wash once with serum-free medium or HBSS.
      2. Observe the cells using a fluorescence microscope with a red excitation filter set for Cy5.
   (b) For analysis by flow cytometry:
      1. Wash once with PBS and add 0.25% trypsin-EDTA solution to detach the cells from the multi-well plate.
      2. Dilute the 0.25% trypsin-EDTA solution with PBS on ice and centrifuge the cell suspension at 500× *g* for 5 min (do not neutralize trypsin with serum).
      3. Discard the supernatant and resuspend the cells in PBS.
      4. Filter the cell suspension through a cell strainer (40 μm nylon mesh) to remove debris.
      5. Analyze the sample using a flow cytometer with an Allophycocyanin (APC) filter.

3.2 Iron Probe for Measuring LIP

The LIP is a redox-active pool of weakly chelated iron that includes both Fe^{2+} and Fe^{3+} irons [8]. LIP size is tightly regulated and plays a critical role in cellular iron transport, iron-regulatory gene expression, iron-containing protein activity, and Fenton reaction catalysis [9]. Changes in intracellular LIP levels can contribute to various cellular injuries, including ferroptosis. Several fluorescent probes have been developed to indicate changes in LIP levels, such as calcein-acetoxymethyl (AM) ester, Phen Green SK, and Phen Green FL [20]. These probes are quenched by iron chelation, and changes in their fluorescent signals can be detected and quantified using a fluorescent microscope or flow cytometry. A novel type of iron probe based on fluorescence resonance energy transfer (FRET) technology has been designed, including FRET iron probe 1 (FIP-1) [21] and DRhFe [22]. FIP-1 can be cleaved by Fe^{2+} in LIP, resulting in decreased FRET from donor fluorophore to acceptor fluorophore. DRhFe is a Fe^{3+}-specific FRET probe with a reversible response. Ratiometric fluorescence imaging is used to detect changes in Fe^{2+} or Fe^{3+} levels in LIP. These probes can be applied to the ferroptosis model to monitor LIP in living cells (*see* **Note 4**).

3.2.1 Calcein-AM Assay

Phen Green SK or Phen Green FL assays follow a protocol similar to that of calcein-AM.

1. Plate PANC1 or HepG2 cells in a glass bottom dish at a satisfactory density and culture them overnight at 37 °C.

2. Remove the culture medium from the dish and rinse twice gently with warm HBSS.
3. For calcein-AM: Add 50 μL of DMSO to 1 vial (50 μg) of calcein-AM to prepare a 1 mM calcein-AM stock solution. For Phen Green SK or Phen Green FL: Add 50 μL of DMSO to 1 vial (100 μg) of Phen Green SK or Phen Green FL to prepare a 1 mM stock solution.
4. Proceed differently for (a) fluorescent microscopy observation or (b) flow cytometry analysis:
   (a) For observation by fluorescent microscopy:
      1. Dilute 1 mM calcein-AM stock solution with HBSS to prepare a 2 μM calcein-AM working solution.
      2. Add the calcein-AM working solution to the glass bottom dish containing PANC1 or HepG2 cells and incubate for 30 min at 37 °C in the dark.
      3. Gently remove the calcein-AM solution and wash the cells once with warm HBSS.
      4. Record the baseline green fluorescence of the cells using a fluorescence microscope.
      5. Treat the cells with an iron chelator (20 μM DFO) or supplement the cells with a Fe^{2+} (50 μM ferrous ammonium sulfate) or Fe^{3+} compound (50 μM $FeCl_3$) for the desired time.
      6. Record the changes in green fluorescence intensity by observing the cells with a fluorescence microscope.
   (b) For analysis by flow cytometry:
      1. Dilute 1 mM calcein-AM stock solution with DMSO to prepare 50 μM calcein-AM working solution.
      2. Add calcein-AM working solution to the dish with a final concentration of 100 nM and incubate for 15–20 min at 37 °C in the dark.
      3. Treat the cells with an iron chelator (20 μM DFO) or supplement the cells with a Fe^{2+} (50 μM ferrous ammonium sulfate) or Fe^{3+} compound (50 μM $FeCl_3$) for the desired time.
      4. Wash the cells with warm PBS and trypsinize them. Centrifuge the cell suspension at 300× *g* and resuspend the cells in warm HBSS.
      5. Analyze the calcein fluorescence of samples by flow cytometry (emission at 517 nm and excitation at 488 nm). The relative size of LIP can be determined by calculating the difference in fluorescence intensity between chelator-untreated and chelator-treated cells.

3.2.2 FIP-1 Assay

The DRhFe assay follows a protocol similar to that of FIP-1 [21].

1. Plate PANC1 or HepG2 cells in a glass bottom dish at a satisfactory density and culture overnight at 37 °C.
2. Remove the culture medium from the dish and rinse twice gently with HBSS.
3. Treat cells with an iron chelator (20 μM DFO) or supplement cells with Fe^{2+} compound (50 μM ferrous ammonium sulfate) for desired time. In DRhFe assay, use Fe^{3+} compound (50 μM $FeCl_3$) instead.
4. Dilute 5 mM FIP-1 stock solution with HBSS to prepare a 10 μM FIP-1 working solution.
5. Add the FIP-1 working solution to the dish and incubate for 90 min at 37 °C in the dark (*see* **Note 5**).
6. Wash twice with HBSS.
7. Observe the cells with a fluorescence microscope to record changes in green fluorescence intensity, with emission at 515 nm and excitation at 488 nm.

4 Notes

1. If the fluorescent signal is too weak to detect, it is recommended to include a known positive control to confirm the results. Alternatively, increasing the concentration of the dye or extending the incubation time may improve the signal strength. It is important to avoid neutralizing trypsin with serum, as the dye may react with Fe^{2+} in the serum prior to reacting with intracellular Fe^{2+}.
2. DMSO is often used as a control in assays because it is a commonly used solvent that does not interact with the sample or assay components. Adding DMSO to the control samples can help identify any nonspecific effects or artifacts in the assay. DMSO is also used as a vehicle to dissolve certain compounds that may be insoluble in water or other solvents. By comparing the results of the DMSO control to the experimental samples, researchers can determine whether any observed effects are due to the compound being tested or are simply a result of the DMSO solvent. It is important to note that DMSO can have some biological effects at high concentrations, so it is important to use a concentration that is not toxic to cells or otherwise interfere with the assay.
3. Iron chelators are compounds that can bind to iron ions and prevent them from participating in chemical reactions. There are several types of iron chelators commonly used in research and medicine. Researchers may select different iron chelators as

a control in iron-detecting probe assays. Some common types of iron chelators include:

(a) Deferoxamine (DFO): a naturally occurring siderophore that binds iron with high affinity and is used to treat iron overload in conditions such as thalassemia and sickle cell disease.

(b) Deferiprone (DFP): an orally active iron chelator used to treat iron overload in patients with thalassemia and other conditions.

(c) Deferasirox (DFX): an orally active iron chelator used to treat chronic iron overload in patients with thalassemia and other conditions.

(d) EDTA (ethylene-diamine-tetraacetic acid): a chelating agent that can bind to various metal ions, including iron.

(e) EGTA (ethylene glycol-bis(2-aminoethylether)-N,N,N′,N′-tetraacetic acid): a chelating agent that is selective for calcium ions but can also bind to iron.

(f) 2,2′-Dipyridyl (DIP): a small molecule chelator that binds specifically to Fe^{2+} ions.

(g) Bathophenanthroline disulfonate (BPS): a fluorescent iron chelator that can bind specifically to Fe^{2+} ions.

4. For fluorescent observation, use glass bottom dishes to allow for better visualization. To avoid interference from ionic iron in serum, use serum-free medium or HBSS. If there is no change in the fluorescent signal, it is important to verify that the correct Fe^{2+} or Fe^{3+} chelator or supplement compound was used with the specific iron probe. Additionally, ensure that the fluorescence detection filter is properly selected and functioning.

5. FIP-1 is sensitive to changes in temperature. Therefore, it is important to maintain the temperature of the assay solutions to ensure the accuracy of the results. Additionally, the probe is sensitive to light, so it is recommended to perform the assay in the dark or low-light conditions.

Acknowledgments

Research by D.T. and R.K. was supported by grants from the National Institutes of Health (R01CA160417, R01CA229275, and R01CA211070).

References

1. Abbaspour N, Hurrell R, Kelishadi R (2014) Review on iron and its importance for human health. J Res Med Sci 19:164–174
2. Lieu PT, Heiskala M, Peterson PA et al (2001) The roles of iron in health and disease. Mol Asp Med 22:1–87
3. Harigae H (2018) Iron metabolism and related diseases: an overview. Int J Hematol 107:5–6
4. Torti SV, Torti FM (2013) Iron and cancer: more ore to be mined. Nat Rev Cancer 13: 342–355
5. Anderson GJ (2007) Mechanisms of iron loading and toxicity. Am J Hematol 82:1128–1131
6. Dixon SJ, Stockwell BR (2014) The role of iron and reactive oxygen species in cell death. Nat Chem Biol 10:9–17
7. Galaris D, Barbouti A, Pantopoulos K (2019) Iron homeostasis and oxidative stress: an intimate relationship. Biochim Biophys Acta, Mol Cell Res 1866:118535
8. Kakhlon O, Cabantchik ZI (2002) The labile iron pool: characterization, measurement, and participation in cellular processes(1). Free Radic Biol Med 33:1037–1046
9. Kruszewski M (2003) Labile iron pool: the main determinant of cellular response to oxidative stress. Mutat Res 531:81–92
10. Chen X, Li J, Kang R et al (2021) Ferroptosis: machinery and regulation. Autophagy 17: 2054–2081
11. Liu J, Kang R, Tang D (2022) Signaling pathways and defense mechanisms of ferroptosis. FEBS J 289:7038–7050
12. Xie Y, Hou W, Song X et al (2016) Ferroptosis: process and function. Cell Death Differ 23: 369–379
13. Chen X, Kang R, Kroemer G et al (2021) Targeting ferroptosis in pancreatic cancer: a double-edged sword. Trends Cancer 7:891–901
14. Dixon SJ, Lemberg KM, Lamprecht MR et al (2012) Ferroptosis: an iron-dependent form of nonapoptotic cell death. Cell 149:1060–1072
15. Hou W, Xie Y, Song X et al (2016) Autophagy promotes ferroptosis by degradation of ferritin. Autophagy 12:1425–1428
16. Liu J, Song X, Kuang F et al (2021) NUPR1 is a critical repressor of ferroptosis. Nat Commun 12:647
17. Tang D, Kroemer G, Kang R (2023) Ferroptosis in hepatocellular carcinoma: from bench to bedside. Hepatology. https://doi.org/10.1097/HEP.0000000000000390. Publish Ahead of Print
18. Hirayama T, Nagasawa H (2017) Chemical tools for detecting Fe ions. J Clin Biochem Nutr 60:39–48
19. Yu Y, Xie Y, Cao L et al (2015) The ferroptosis inducer erastin enhances sensitivity of acute myeloid leukemia cells to chemotherapeutic agents. Mol Cell Oncol 2:e1054549
20. Petrat F, Rauen U, de Groot H (1999) Determination of the chelatable iron pool of isolated rat hepatocytes by digital fluorescence microscopy using the fluorescent probe, phen green SK. Hepatology 29:1171–1179
21. Aron AT, Loehr MO, Bogena J et al (2016) An Endoperoxide reactivity-based FRET probe for Ratiometric fluorescence imaging of labile iron pools in living cells. J Am Chem Soc 138: 14338–14346
22. Gao J, He Y, Chen Y et al (2020) Reversible FRET fluorescent probe for Ratiometric tracking of endogenous Fe(3+) in Ferroptosis. Inorg Chem 59:10920–10927

Chapter 2

Stratifying Ferroptosis Sensitivity in Cells and Tissues with PALP

Fengxiang Wang, Nathchar Naowarojna, and Yilong Zou

Abstract

Ferroptosis is emerging as a promising strategy for suppressing multiple types of human cancers. Rapid and accurate assessment of the relative sensitivity to ferroptosis in biological samples will accelerate the development of ferroptosis-targeted therapies. We previously demonstrated that photochemical activation of membrane lipid peroxidation (PALP) that uses high-power lasers to induce localized polyunsaturated fatty acyl (PUFA)-lipid peroxidation can efficiently report ferroptosis sensitivity in live cells and tissues in situ. Here, we describe the experimental details for PALP analysis, including preparation of tissue sections, preparation of fluorescent lipid peroxidation reporter, sample staining, lipid peroxidation induced by laser source, and data processing. We envision predicting the relative sensitivity to ferroptosis of cellular and tissue samples is potentially useful for basic research and clinical investigations.

Key words Ferroptosis sensitivity, Lipid peroxidation, Tissue in situ imaging, Live-cell imaging, Tumor, Polyunsaturated-lipids

1 Introduction

Ferroptosis is a regulated cell death program marked by iron-dependent accumulation of lipid hydroperoxides [1]. Ferroptosis is implicated in the development of various diseases including neurodegeneration, ischemia/reperfusion-induced damages in the liver, kidney, heart, and brain, and hemolysis [2]. Various human cancers including clear-cell carcinomas from the kidney and ovary, pancreatic cancer, diffused large B-cell lymphoma, hepatocellular carcinoma, colorectal cancer, and therapy-resistant cancer cell "persisters" are susceptible to ferroptosis induction via pharmacological or genetic approaches [3–8]. Hence, inducing ferroptosis in cancer cells has emerged as a promising anti-cancer strategy [9]. However, rapidly stratifying cancer patients for their likelihood to respond to ferroptosis-inducing therapeutic agents remains a major challenge in developing ferroptosis-targeted anti-cancer

Guido Kroemer and Daolin Tang (eds.), *Ferroptosis: Methods and Protocols*, Methods in Molecular Biology, vol. 2712,
https://doi.org/10.1007/978-1-0716-3433-2_2,

treatments. This bottleneck is due to a lack of broadly validated biomarkers and tools to predict the ferroptosis susceptibility of a biological sample as well as to report the effectiveness of treatment. This issue is magnified by the heterogeneity in ferroptosis sensitivity and metabolism across cancer lineages, patients, and even different tumor lesions from the same patient.

In ferroptosis, the regulation of lipid metabolism plays a key role in dictating cellular susceptibility to ferroptosis. Cellular membrane lipids, specifically polyunsaturated phospholipids (PUFA-PLs), are highly susceptible to lipid peroxidation due to the presence of bis-allylic H atoms [10, 11]. Hence, the abundance of cellular PUFA-PLs is a major determinant of cellular sensitivity to ferroptosis induction. The prominent role of PUFA-PLs in shaping ferroptosis sensitivity is supported by the general requirement of acyl-CoA synthetase long-chain family member 4 (ACSL4), a fatty acid activator with high selectivity toward PUFAs, in ferroptosis occurring in various contexts [12, 13]. Here, we describe a detailed protocol of an imaging approach, photochemical activation of membrane lipid peroxidation (PALP), aimed to track lipid peroxidation dynamics, assess polyunsaturated phospholipid levels, and stratify ferroptosis sensitivity of measured samples including live cells and tissues [14]. This protocol uses BODIPY™ 581/591 C11 (BODIPY-C11, Life Technologies, D3861) as the reporter of lipid peroxidation. BODIPY-C11 can be used to detect reactive oxygen species (ROS) in lipophilic environments such as cellular membranes. Oxidation of the polyunsaturated butadienyl structure in BODIPY-C11 results in a shift of fluorescence emission peak from ~590 nm to ~510 nm [15]. PALP stimulation can induce strong oxidative BODIPY-C11 signals in ferroptosis-susceptible samples enabling the stratification of ferroptosis sensitivity in situ.

2 Materials

2.1 Cell Culture and Tissue Sectioning

1. Penicillin/streptomycin.
2. Fetal bovine serum.
3. Phosphate buffer saline.
4. RPMI 1640 medium.
5. DMEM medium.
6. 35 mm glass bottom imaging dish.
7. Microscope slide.
8. Microscope cover slide.

2.2 Preparation of Fluorescent Lipid Peroxidation Reporter

1. Prepare stock BODIPY-C11 solution at 5 mM (1000×) concentration in DMSO. Aliquot the stock solution into small vials and store at −20 °C to avoid repetitive freeze-thaw. This reagent should be shielded from light.
2. Dilute BODIPY-C11 at 5 μM (1×) of working concentration. For live-cell imaging, use complete culture medium to dilute. To analyze tissue sections, use fetal bovine serum or phosphate buffer saline (PBS, pH 7.4) as the dilution solvent. Keep the working solution at room temperature and shielded from light prior to use.

2.3 Measurement of Lipid Peroxidation

2.3.1 Required Software

1. ZEN 3.3 blue edition (Zeiss Microscope).
2. Nikon NIS Elements software (Nikon NIS Microscope).
3. ImageJ 1.52P (Fiji).
4. GraphPad Prism 9.0 (GraphPad Software Inc).

2.3.2 Required Equipment

1. Cryostat tissue sectioning equipment (Leica, CM1950).
2. A1R HD25 confocal microscope system (Nikon).
3. LSM 800 confocal microscope system with Airyscan (Zeiss).

3 Methods

3.1 Collecting and Snap-Freezing the Tissue Samples

1. Euthanize the mouse by cervical dislocation and dissect the tissue.
2. Rapidly wash the tissue samples with deionized water to remove residual blood.
3. Snap-freeze tissue (*see* Fig. 1a).
   (a) Fold a piece of aluminum foil into a boat shape.
   (b) Place tissue inside the aluminum boat and float it on liquid nitrogen.
   (c) After 5 s, gently press down the aluminum boat with forceps and allow a small amount of liquid nitrogen to flow into the boat. Repeat until the tissue is completely submerged in liquid nitrogen. This step should be finished within 1 min.
   (d) Submerge the tissue in liquid nitrogen for 15 s, remove the residual liquid nitrogen and wrap tissue with aluminum foil.
   (e) Transfer the tissue to −80 °C using dry ice and equilibrate for over 1 h prior to tissue sectioning.

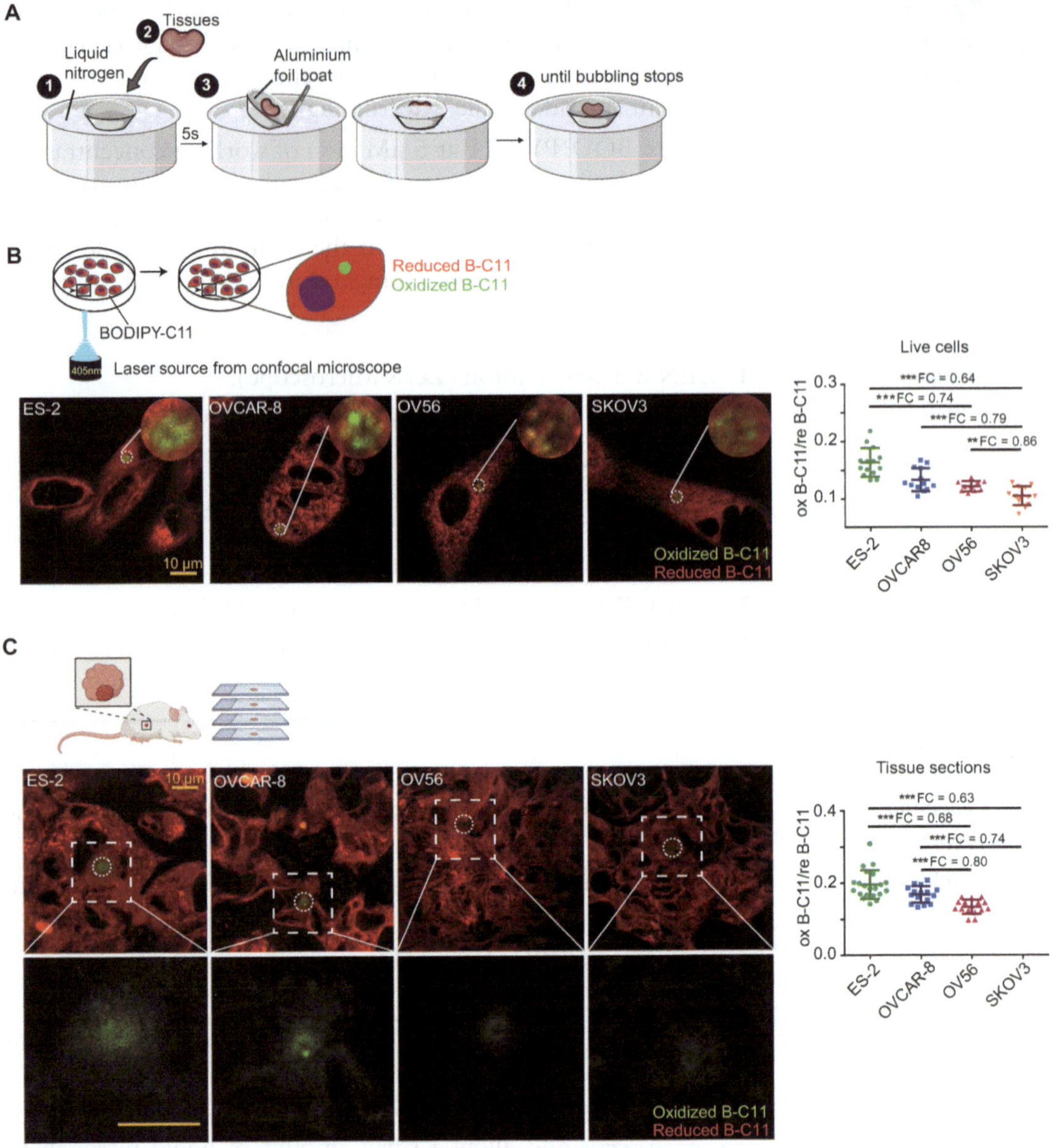

Fig. 1 Photochemical activation of membrane lipid peroxidation (**a**) Scheme for snap-freezing tissues in liquid nitrogen. (**b**). PALP-induced fluorescence in live cells. Graphical scheme summarizing the PALP technique. Representative fluorescent images showing the PALP-induced fluorescence of human ovarian ES-2, OVCAR-8, OV56, and SKOV3 cells. Scatter plots showing the quantifications of the PALP-induced fluorescence intensities by ox B-C11/re B-C11 in human cancer cells. (**c**) PALP-induced fluorescence in xenograft tumors. Graphical scheme describing the sample preparation procedures of PALP technique for tissue sections. Representative fluorescent images showing the PALP-induced fluorescence of xenograft tumor sections including human ovarian ES-2, OVCAR-8, OV56, and SKOV3 cells. Scatter plots showing the quantifications of the PALP-induced fluorescence intensities by ox B-C11/re B-C11 in xenograft tumor sections. Lines and error bars, mean $\pm$ s.d. Two-tailed unpaired T-test, ***, $p < 0.001$. *FC* fold-change. Green, oxidized BODIPY-C11 signal, ox B-C11; red, reduced BODIPY-C11 signal, re B-C11. The results presented in this protocol are reprinted from Wang et al. [14, 16] with permission from Elsevier

3.2 Cryosectioning Tissues (see Note 1)

1. Equilibrate the frozen tissue in −20 °C for 30 min prior to sectioning (*see* **Note 2**).
2. Transfer the tissue to the cryostat chamber and affix the tissue to the specimen disc using ultrapure water.
3. Set the optimal sectioning temperature, which typically is −20 ± 3 °C.
4. Section the tissue to the desired thickness between 5 and 15 μm, which typically is 10 μm. Thaw mount sections onto glass slides (*see* **Note 3**).

3.3 Pre-staining Sample with BODIPY-C11 (see Notes 4–6)

For live-cell imaging, start from this step. Keep samples shielded from light after BODIPY-C11 staining.

1. For live-cell imaging (*see* Fig. 1b):
   (a) Replace the culture medium with the BODIPY-C11 working solution prepared with a complete culture medium.
   (b) Incubate cells in a 5% CO_2, 37 °C incubator for 30 min.
2. For tissue imaging (*see* Fig. 1c):
   (a) Equilibrate the tissue section slide to room temperature.
   (b) Draw a circle around the tissue using a hydrophobic pen.
   (c) Apply 100 μL BODIPY-C11 working solution onto the sample surface.
   (d) Cover the samples with coverslips.
   (e) Incubate the tissue section at 37 °C for 30 min.

3.4 Inducing Lipid Peroxidation and Stratifying Ferroptosis Sensitivity (see Notes 6 and 7)

1. Place the BODIPY-C11 conjugated sample to the microscope stage and adjust the focal plane (*see* **Note 8**).
2. Adjust the intensity of 488 nm and 561 nm excitation laser sources and detector voltage to avoid overexposure.
3. Acquire the pre-laser stimulation fluorescent signals of pre-oxidized (Ox_{pre}) and pre-reduced (Re_{pre}) images from 488 nm (FITC) and 561 nm (TRITC) excitation light channels (*see* Fig. 1b).
4. Select a small region of interest (ROI) to focus the energy from the laser to enable lipid peroxidation induction, usually around 2–4 μm^2.
5. Use the 405 nm solid-state laser source from a confocal microscope to excite the selected region and induce lipid peroxidation. Typically use the laser stimulating for 5 repeats of 200-ms interval using Nikon A1R confocal microscope (*see* **Note 9**).

6. Immediately acquire post-laser stimulation fluorescent signals of post-oxidized (Ox_{post}) and post-reduced (Re_{post}) BODIPY-C11 images from 488 nm (FITC) and 561 nm (TRITC) excitation light channels (*see* **Note 10**, Fig. 1b).

3.5 Data Processing: Quantify PALP Signals on Regions of Interest (see Notes 11 and 12)

1. Analyze the fluorescent signal of the measured region using image processing softwares (*see* **Note 13**).
   (a) Import the ROI to locate stimulated regions.
   (b) Analyze the gray intensity of pre-oxidized (Ox_{pre}), pre-reduced (Re_{pre}), and post-oxidized (Ox_{post}) images.
2. Normalize the lipid peroxidation signals induced by laser stimulation across different replicates and groups (*see* Fig. 1b, c):
   (a) To normalize the different absorption and integration efficiency of BODIPY-C11: Ox_{post}/Re_{pre}, i.e., the oxidized BODIPY-C11 intensity post-stimulation ($\lambda_{Ex/Em}$: 488/510 nm) divided by the reduced BODIPY-C11 fluorescence pre-stimulation ($\lambda_{Ex/Em}$: 581/591 nm).
   (b) To exclude the background signals of oxidative lipids: $(Ox_{post} - Ox_{pre})/Re_{pre}$, i.e., the oxidized BODIPY-C11 intensity post-stimulation subtract pre-stimulation ($\lambda_{Ex/Em}$: 488/510 nm) and divided by the reduced BODIPY-C11 fluorescence pre-stimulation ($\lambda_{Ex/Em}$: 581/591 nm).

4 Notes

1. During cryostat sectioning, tissues should be kept frozen at a steady temperature and avoid temperature fluctuations (at − 20 °C to minimize metabolic changes)
2. To embed thin or fragile tissues such as skin, testis, and ovary, use 10% gelatin in 5% carboxymethylcellulose (CMC) as the embedding medium [17]
3. For long-term storage, keep the mounted tissue section slides at −80 °C without frequent freeze-thaw cycles and minimize light exposure
4. After adding BODIPY-C11, samples need to be incubated at 37 °C for staining and are often kept at room temperature (22 ± 3 °C) for imaging. Do not freeze the sample sections after BODIPY-C11 administration
5. Even distribution of the BODIPY-C11 dye on the sample sections is important for the validity of intra-sample comparisons across different sample regions
6. In principle, BODIPY-C11 could be substituted by other fluorescent lipid peroxidation reporters.

7. Other brands and models of confocal microscope could be employed to induce lipid peroxidation under adjusted laser power and frequency. However, different models of solid-state laser source have different power outputs and varying lipid peroxidation induction efficiency. Hence, quantitative features of PALP-induced signals from different microscope systems may not be directly compared.
8. To obtain high spatial resolution, laser administration via high-magnification objective lenses, such as 60×, or 63× lenses commonly equipped on confocal microscopes is recommended. The lens choices can be adjusted based on the user's preferences.
9. The stimulation time and frequency for different microscope models should be adjusted to enable lipid peroxidation induction while avoiding quenching the BODIPY-C11 dye.
10. We recommend stimulating multiple regions of one replicate sample and obtaining a comprehensive assessment of the ferroptosis sensitivity using at least three biological replicates.
11. BODIPY-C11 is a lipophilic fluorescent molecule that penetrates cellular membranes. Different types of cells exhibit different lipid content and varying ranges of BODIPY-C11 penetration, resulting in variations in the intensity and distribution of reduced BODIPY-C11 signals. To assess ferroptosis sensitivity, we recommend using the ratio between oxidized and reduced BODIPY-C11 (Ox_{post}/Re_{pre}), rather than the post-oxidized BODIPY-C11 signal only. To exclude the substantial basal level of pre-oxidized BODIPY-C11 signal, consider normalizing lipid peroxidation signal by $(Ox_{post} - Ox_{pre})/Re_{pre}$.
12. Post-oxidized BODIPY-C11 signals can only be excited from regions where the BODIPY-C11 molecule effectively penetrates and stains, for instance, the endoplasmic reticulum membranes in most cells. We recommend using pre-reduced BODIPY-C11 signal distribution to guide the stimulation targeting regions. For live cells, cellular structures that exhibit strong pre-reduced BODIPY-C11 signals are appropriate candidate regions for PALP analyses.
13. In this Method, imaging analysis was acquired by Nikon NIS Elements software for Nikon microscope, Metamorph software for Andor confocal microscope, and ZEN blue edition for Zeiss microscope. Softwares usage can be adjusted according to the users' microscope preferences. Downstream fluorescent image analysis can be performed using Fiji ImageJ or equivalent.
14. Resistance to ferroptosis induction could occur in tumors in vivo [18]. Due to the potential complexities in the

mechanisms mediating ferroptosis evasion, it remains to be tested whether PALP, assessing content of polyunsaturated fatty acyl-lipids, can be used to assess and predict acquired resistance to ferroptosis induction in tumors in vivo.

15. PALP technique is a semi-quantitative strategy to stratify ferroptosis sensitivity among groups of specimens. Considering the different characteristics of different sample types, sectioning instruments and conditions, etc., we recommend the users to optimize and standardize their protocols prior to large-scale assessment of their samples for improved reliability in inter-sample comparisons.

Declaration of Interests

Y.Z. is a consultant of Keen Therapeutics. The remaining authors declare no competing interests. A patent describing the utility of PALP in tissue samples was filed by Y.Z., F.W., N.N., et al.

References

1. Dixon SJ, Kathryn ML, Michael RL et al (2012) Ferroptosis: an iron-dependent form of nonapoptotic cell death. Cell 149(5): 1060–1072
2. Zou Y, Stuart LS (2020) Progress in understanding ferroptosis and challenges in its targeting for therapeutic benefit. Cell Chem Biol 27(4):463–471
3. Zou Y, Michael JP, Amy AD et al (2019) A GPX4-dependent cancer cell state underlies the clear-cell morphology and confers sensitivity to ferroptosis. Nat Commun 10(1):1617
4. Badgley MA, Daniel MK, Maurer HC et al (2020) Cysteine depletion induces pancreatic tumor ferroptosis in mice. Science 368(6486): 85–89
5. Yang WS, Rohitha S, Matthew EW et al (2014) Regulation of ferroptotic cancer cell death by GPX4. Cell 156(1–2):317–331
6. Sun X, Ou Z, Chen R et al (2016) Activation of the p62-Keap1-NRF2 pathway protects against ferroptosis in hepatocellular carcinoma cells. Hepatology 63(1):173–184
7. Singhal R, Sreedhar RM, Nupur KD et al (2021) HIF-2α activation potentiates oxidative cell death in colorectal cancers by increasing cellular iron. J Clin Invest 131(12). https://doi.org/10.1172/JCI143691
8. Viswanathan VS, Matthew JR, Harshil DD et al (2017) Dependency of a therapy-resistant state of cancer cells on a lipid peroxidase pathway. Nature 547(7664):453–457
9. Chen X, Kang R, Kroemer G, Tang D (2021). Broadening horizons: the role of ferroptosis in cancer. Nat Rev Clin Oncol 18(5):280–296.
10. Conrad M, Derek AP (2019) The chemical basis of ferroptosis. Nat Chem Biol 15(12): 1137–1147
11. Yang WS, Katherine JK, Michael MG et al (2016) Peroxidation of polyunsaturated fatty acids by lipoxygenases drives ferroptosis. Proc Natl Acad Sci U S A 113(34):E4966–E4975
12. Kagan VE, Gaowei M, Feng Q et al (2017) Oxidized arachidonic and adrenic PEs navigate cells to ferroptosis. Nat Chem Biol 13(1): 81–90
13. Doll S, Bettina P, Yulia YT et al (2017) ACSL4 dictates ferroptosis sensitivity by shaping cellular lipid composition. Nat Chem Biol 13(1): 91–98
14. Wang F, Emily TG, Nathchar N et al (2022) PALP: a rapid imaging technique for stratifying ferroptosis sensitivity in normal and tumor tissues in situ. Cell Chem Biol 29(1): 157–170
15. Drummen GPC, Lydia CM, Jos AFO et al (2002) C11-BODIPY581/591, an oxidation-sensitive fluorescent lipid peroxidation probe: (micro)spectroscopic characterization and

validation of methodology. Free Radic Biol Med 33(4):473–490

16. Wang F, Nathchar N, Zou Y (2022) Stratifying ferroptosis sensitivity in cells and mouse tissues by photochemical activation of lipid peroxidation and fluorescent imaging. STAR Protoc 3(2)
17. Nelson KA, Gabrielle JD, John WF et al (2013) Optimization of whole-body zebrafish sectioning methods for mass spectrometry imaging. J Biomol Tech 24(3):119–127
18. Zou Y, Whitney SH, Emily LR et al (2020) Plasticity of ether lipids promotes ferroptosis susceptibility and evasion. Nature 585(7826): 603–608

Chapter 3

ChIP and ChIRP Assays in Ferroptosis

Zuli Wang, Tania Tao, and Yongguang Tao

Abstract

Ferroptosis is characterized by the accumulation of lipid peroxidation driven by iron. As a regulated cell death, ferroptosis plays a critical role in various diseases and exhibits great therapeutic potentials. However, the mechanisms underlying ferroptosis, including its occurrence, execution, and regulation, remain poorly understood, which is necessary for developing effective therapeutic strategies. In this chapter, we summarize chromatin immunoprecipitation (ChIP) assay for the research of proteins-chromatin interactions. Moreover, Chromatin Isolation by RNA Purification (ChIRP) trial is introduced to investigate the interactions between lncRNA and chromatin. The application of ChIP and ChIRP is expected to explore the transcription and epigenetic regulation of ferroptosis deeply for therapeutic benefits.

Key words ChIP, ChIRP, Ferroptosis

1 ChIP Assay in Ferroptosis

1.1 Introduction

Chromatin structures, consisting of DNA, proteins, and RNA, are always in dynamic state in response to intra- and extracellular stimuli via controlling the transcription of related genes. Chromatin immunoprecipitations (ChIP) is a key technology widely used to demonstrate the natural interactions between proteins and DNA in vivo and further elucidate these processes [1]. For example, researchers usually apply ChIP assay to validate the binding region of defined transcription factor or histone to candidate DNA sequence [2]. For the ChIP protocol, cellular protein-DNA complexes are fixed and cross-linked using formaldehyde. Then, after sonication or enzymatic digestion to shear chromatin, the products are immunoprecipitated and captured by antibody. Subsequently, isolated and purified DNAs are usually detected by quantitative polymerase chain reaction (qPCR) [3, 4].

Given the limitation of ChIP-qPCR to defined proteins and specific DNA, ChIP combined with genome-wide profiling approaches including microarray hybridization (ChIP-on-chip) or massively parallel sequencing (ChIP-seq) not only promote the

Guido Kroemer and Daolin Tang (eds.), *Ferroptosis: Methods and Protocols*, Methods in Molecular Biology, vol. 2712,
https://doi.org/10.1007/978-1-0716-3433-2_3,

discoveries of transcriptional control and epigenetic regulation, but also help us easily understand the organization and function of the complex genomes [5]. Moreover, compared to ChIP-on-chip technology, ChIP-seq is more commonly used throughout scientific researches for its high resolution and wide detection range [6, 7]. Recently, the Drop-ChIP, namely single-cell ChIP-seq, which is developed by Bradley E. Bernstein professor group aims at analyzing the high-quality chromatin state of single cell and identifying a chromatin signatures spectrum for cell population [8].

ChIP assay is widely used transcriptional control and epigenetic regulation of ferroptosis-related genes. For example, transcription factors ATF3, BACH1, and TP53 are demonstrated to enhance the sensitivity of cells to ferroptosis via downregulation of SLC7A11 [9]. Conversely, YAP-1 and SP-1 upregulate the expression of ACSL4 through transcription, leading to ferroptotic aggravation [10]. CFP1 epigenetically activates the expression of lncRNA P53RRA, which enhances the interaction with p53-interacting protein G3BP1, promoting TP53-mediated apoptosis and ferroptosis [11, 12]. Moreover, lymphoid-specific helicase (LSH) interacts with WDR76 to increase the level of H3K4Me3, an active chromatin marker, at the promoters of ferroptosis-related genes SCD1 and FADS2 [13]. Overall, ChIP assay greatly contributes to the exploration of the mechanisms of ferroptosis.

1.2 Materials

1.2.1 Materials and Equipments

1. 10 cm dish.
2. Rotator.
3. Sonicator.
4. 15 mL centrifuge tube.
5. Centrifuge.
6. 1.5 mL centrifuge tube.
7. Magnetic separator.
8. Real-Time PCR Detect System.

1.2.2 Reagents and Buffers

1. PBS buffer (137 mM NaCl, 2.7 mM KCl, 8 mM Na_2HPO_4, 1.8 mM KH_2PO_4).
2. 37% formaldehyde (Sigma-Aldrich).
3. 1.25 M glycine (Sangon Biotech).
4. Protease inhibitor cocktail (Sigma-Aldrich).
5. Cell lysis buffer (Cell Signaling Technology).
6. IgG antibody (Abcam).
7. SDS lysis buffer (50 mM Tris pH 8.0, 1% SDS).
8. BCA (Thermo Fisher Scientific).
9. PCR purification kit (Sangon Biotech).

10. Protein G Dynabeads (Invitrogen).
11. Low salt wash buffer (0.1% SDS, 1% TritonX-100, 2 mM EDTA, 20 mM Tris-HCl pH 8.0, 150 mM NaCl).
12. High salt wash buffer (0.1% SDS, 1% TritonX-100, 2 mM EDTA, 20 mM Tris-HCl pH 8.0, 500 mM NaCl).
13. LiCl buffer (0.25 M LiCl, 1% NP-40, 1% SDS, 1 mM EDTA, 10 mM Tris-HCl).
14. Chelex-100 (Sigma-Aldrich).
15. FastStart Universal SYBR Green Master (Roche).

1.3 Methods

A modified ChIP protocol from MERCK (https://www.sigmaaldrich.cn/CN/zh/product/mm/17371) is described in detail as follows:

1. Rinse the 10 cm dish-cultured cells with 1× PBS.
2. Add 10 mL medium and 270 μL 37% formaldehyde into the dish. Shake it gently for 10 min at room temperature.
3. Add 1 mL 1.25 M glycine to terminate the reaction. Shake cells for 5 min and rinse cells twice with ice-cold 10 mL PBS.
4. Add 1 mL PBS and protease inhibitor cocktail to the culture dish.
5. Collect cells into a 15 mL centrifuge tube and spin at 500 g for 5 min at 4 °C.
6. Resuspend the cell pellet with 5 mL cell lysis buffer. Incubate on ice for 5 min and spin at 500 g for 5 min at 4 °C.
7. Resuspend the cell pellet with 1 mL cell lysis buffer and transfer to 1.5 mL centrifuge tube. Spin at 500 g for 5 min at 4 °C.
8. Add 300 μL SDS lysis buffer containing cocktail to resuspend the pellet, and incubate on ice for 30 min.
9. The solution is performed for sonication (100%, 20 sec On, 20 sec Off, 3 min). Spin at 13,000 rpm for 10 min at 4 °C.
10. Transfer the supernatant into a new 1.5 mL centrifuge tube. Protein concentration is measured by the BCA method. Set 300 μg for each group including Input group, IgG group, and experimental group.
11. Input group is performed with PCR purification kit, and elute with 200 μL TE buffer. Check DNA size on 1% agarose gel until the DNA fragment is 150–750 bp.
12. Add cell lysis buffer to 500 μL in IgG group and experimental group respectively, then add 10 μL Protein G Dynabeads, rotate at 4 °C for 1 hr.; add cell lysis buffer to 200 μL in the Input group.

13. Transfer the supernatant into a new 1.5 mL centrifuge tube. Add 2 μg primary antibody to the experimental group and 2 μL IgG antibody to the IgG group, and rotate at 4 °C for overnight.
14. Add 20 μL Protein G Dynabeads into each tube, and rotate for 40 min.
15. Wash the magnetic beads effectively: Low Salt Wash Buffer ×1, High Salt Wash Buffer ×1, LiCl Wash Buffer ×1, TE buffer ×2, and rotate for 10 min.
16. Add 100 μL 10% chelex-100 to each tube, incubate at 99 °C for 10 min. Spin at 13,000 rpm for 2 min at 4 °C, transfer the supernatant into a new centrifuge tube.
17. Resuspend beads with 125 μL TE buffer, spin at 13,000 rpm for 2 min at 4 °C, transfer the supernatant into previous tube and mix well.
18. The products are analyzed by fluorescent quantitative PCR (polymerase chain reaction) reaction using FastStart Universal SYBR Green Master and primers for the promoter region of the target genes.

1.4 Notes [4]

1. The size of sonicated chromatin should be between 150–750 bp and be verified before starting the immunoprecipitation step.
2. When the background of IgG group is too high, one possible reason is that chromatin binds with magnetic beads in a non-specific manner. Thus, we need to incubate the beads with chromatin complexes for 1 h before the antibody is added or increase the times of beads washing before performing DNA extraction step. If the problem remains unsolved, it is better to purchase new magnetic beads and wash buffer.
3. If the recycling rates of DNA are low, we need to increase the amounts of cells or the dose of antibody. The amounts of cells are at least 3×10^6 cells for each sample. Moreover, make sure the magnetic beads or antibody is proper for ChIP. Besides, excessive cross-linking or insufficient cross-linking reaction may reduce the recycling rates.

2 ChIRP Assay in Ferroptosis

2.1 Introduction

Many long noncoding RNAs (lncRNAs) can bind proteins and/or DNA to regulate gene expression, thus involving cell function and fate. For example, lncRNA LINC00336 stabilized by RNA-binding protein ELAVL1 inhibits ferroptosis via increasing the expression of cystathionine-β-synthase (CBS) [14]. However, the concrete regulatory mechanisms of lncRNA greatly remain to be explored. Chromatin Isolation by RNA Purification (ChIRP), as a novel technique for studying lncRNA-chromatin interactions, not only identifies lncRNAs' genomic binding sites but also that of their interaction proteins [15]. The workflow of ChIRP is similar to ChIP assay. Briefly, chromatin is crosslinked to lncRNA in vivo and homogenized by sonication. Subsequently, biotin-labeled oligonucleotides are hybridized to target lncRNA, which are purified using streptavidin magnetic beads. RNA, DNA, or protein is eluted respectively and then subjected to downstream assays [16, 17].

Recent study reports that lncRNA BDNF-AS can recruit WDR5 to affect FBXW7 transcription and then regulate gastric cancer ferroptosis through VDAC3 ubiquitination using ChIRP [18]. Besides, ChIRP trial demonstrates that lncRNA 00618 interacts with LSH and binds to the promoter region of SLC7A11, thus regulating cell ferroptosis [19]. Given the critical epigenetic regulation role of multiple lncRNAs in ferroptosis [20, 21], the proteins and DNA interacting with lncRNAs are worth investigating using ChIRP assay.

2.2 Materials

2.2.1 Materials and Equipments

1. 10 cm dish.
2. Rotator.
3. Liquid nitrogen container.
4. −80 °C ice box.
5. Sonicator.
6. 15 mL centrifuge tube.
7. Centrifuge.
8. 1.5 mL centrifuge tube.
9. Magnetic separator.
10. Real-Time PCR Detect System.

2.2.2 Reagents and Buffers

1. PBS (137 mM NaCl, 2.7 mM KCl, 8 mM Na_2HPO_4, 1.8 mM KH_2PO_4).
2. Trypsin (Gibco).
3. 1% glutaraldehyde solution (0.8 mL 25% glutaraldehyde stock, 19.2 mL PBS).
4. Glycine (Sangon Biotech).
5. Lysis Buffer (50 mM Tris-Cl pH 7.0, 10 mM EDTA, 1% SDS).
6. Protease Inhibitor (Sigma-Aldrich).
7. Superase-in (Invitrogen).
8. DNA Protease K Buffer (100 mM NaCl, 10 mM Tris-Cl pH 8.0, 1 mM EDTA, 0.5% SDS).
9. Protease K (Ambion).

10. PCR purification kit (Sangon Biotech).
11. Hybridization Buffer (750 mM NaCl, 1% SDS, 50 mM Tris-Cl pH 7.0, 1 mM EDTA, 15% formamide).
12. C-1 magnetic beads (Invitrogen).
13. Wash Buffer (0.3 M NaCl, 0.03 M sodium citrate (2× SSC), 0.5% SDS).
14. TRIzol (Invitrogen).
15. miRNeasy mini columns (Qiagen).
16. DNA Elution Buffer (50 mM NaHCO3, 1% SDS).
17. RNases (Sigma-Aldrich).
18. PhOH:Chloroform:Isoamyl (Invitrogen).
19. DNase buffer (100 mM NaCl, 0.1% NP-40).
20. DNase I (Invitrogen).
21. 2× laemmli buffer (Absin).
22. PVDF membrane (Millipore).

2.3 Methods

A modified ChIRP protocol from Journal of Visualized Experiments (https://www.jove.com/t/3912/chromatin-isolation-by-rna-purification-(chirp)) is described in detail as follows:

1. Probe design at https://www.biosearchtech.com/products/rna-fish/.
2. Rinse the 10 cm dish-cultured cells (×2 for one ChIRP sample) with 1× PBS. Trypsinize cells and quench trypsin with 2× volume medium.
3. Collect and transfer cells to 15 mL centrifuge tube. Spin at 800 g for 5 min.
4. Resuspend cells in 10 mL PBS and Spin at 800 g for 5 min.
5. Resuspend cells in 10 mL 1% glutaraldehyde solution. Rotate for 10 min at room temperature.
6. Quench the crosslink reaction with 1 mL 1.25 M glycine and rotate for 5 min.
7. Spin at 2000 g for 3 min. Wash cells with 10 mL ice-cold PBS and spin at 2000 g for 3 min.
8. Freeze the cell pellets in liquid nitrogen quickly and store at −80 °C.
9. Thaw frozen cells at room temperature. Spin at 2000 g for 3 min at 4 °C.
10. Add 10× volume of Lysis Buffer (e.g., 1 mL for 100 mg) with fresh Protease Inhibitor and Superase-in to resuspend cells.
11. Sonicate cells in a 4 °C water bath at setting with 100%, 30 sec On, 45 s Off.

12. Transfer 5 μL lysate to a new 1.5 mL centrifuge tube. Add 90 μL DNA Protease K Buffer and 5 μL Protease K. Mix and incubate for 45 min at 50 °C.
13. Extract DNA with PCR purification kit. Elute DNA in 30 μL Elution Buffer and check DNA size on 1% agarose gel until the DNA fragment is 150–600 bp.
14. Spin sonicated samples at 16100 g for 10 min at 4 °C. Flash-freeze tubes in liquid nitrogen and store at −80 °C.
15. Thaw tubes at room temperature. Remove 10 μL for RNA Input, 10 μL for DNA Input and 10 μL for protein Input. Transfer 1 mL chromatin to 15 mL centrifuge tube, and add 2 mL Hybridization Buffer with Protease Inhibitor and Superase-in.
16. Add 200 pmol probes to 15 mL Hybridized tubes. Mix well and incubate at 37 °C for 4 h with shaking.
17. Resuspend 100 μL C-1 magnetic beads in original volume of supplemented lysis buffer. Incubate at 37 °C for 30 min with shaking.
18. Wash beads with 1 mL Wash Buffer for four times. Resuspend beads with 2 mL Wash Buffer.
19. Remove 100 μL beads for RNA isolation and reserve 900 μL beads for DNA fraction. Remove 1 mL beads for protein elution.
20. Resuspend 100 μL RNA isolation sample in 95 μL RNA Proteinease K Buffer and add 5 μL Proteinease K. Add 85 μL RNA Proteinease K Buffer to 10 μL RNA Input sample and add 5 μL Proteinease K. Incubate these samples at 50 °C for 45 min.
21. Boil these samples for 10 min at 95 °C and chill on ice.
22. Add 500 μL TRIzol. Vortex violently for 10 s and incubate at room temperature for 10 min.
23. Add 100 μL chloroform and vortex violently for 10 s Spin at 16,100 g for 15 min at 4 °C.
24. Remove 400 μL supernatant and add 600 μL 100% ethanol. Spin sample through miRNeasy mini columns. Wash 1× with RWT and 2× with RPE.
25. Elute samples with 30 μL nuclease-free H_2O. Heat for 15 min at 65 °C.
26. Use 1 μL RNA for fluorescent quantitative PCR analysis, often using GAPDH as a negative control, to confirm lncRNA retrieval.
27. Resuspend DNA fraction samples in 150 μL DNA Elution Buffer with RNases. Resuspend DNA Input sample in

140 μL DNA Elution Buffer. Incubate and shake at 37 °C for 30 min with shaking.

28. Add 15 μL Proteinease K to each sample. Incubate and shake at 50 °C for 45 min.
29. Transfer DNA samples to phase-lock gel tubes and dd 300 μL PhOH:Chloroform:Isoamyl to each sample. Shake vigorously for 10 min, and spin down at 16,100 g for 5 min at 4 °C.
30. Remove 300 μL aqueous. Add 3 μL GlycoBlue, 30 μL NaOAc and 900 μL 100% EtOH. Mix well and store at −20 °C overnight.
31. Spin at 16,100 g for 30 min at 4 °C.
32. Add 1 mL 70% EtOH and vortex to mix. Spin at 16,100 g for 5 min. Air dry and resuspend in 30 μL Elution Buffer.
33. Analyze the DNA samples by fluorescent quantitative PCR.
34. Resuspend protein elution samples in 150 μL DNase buffer with cocktail, RNase A and DNase I. Incubate at 37 °C for 30 min.
35. Add 0.2 volume of 2× laemmli buffer, and boil for 5 min.
36. Dot blot to PVDF membrane. Membrane was blotted against antibodies per normal Western protocol.

2.4 Notes [15]

1. To acquire enough chromatin for DNA and protein detection, the amounts of cells are at least 6×10^6 cells for each sample.
2. Check lysate with sonicate every 30 min until the cell lysate is no longer turbid. If the DNA fragment is 150–600 bp, sonication is complete.
3. The whole process must use experimental equipment and reagents without RNase pollution.
4. The protease inhibitor should be added to the lysis buffer before use.
5. The protein samples are mixed with loading buffer without glycerol so as to better imprint to PVDF membrane.

References

1. Nelson JD, Denisenko O, Bomsztyk K (2006) Protocol for the fast chromatin immunoprecipitation (ChIP) method. Nat Protoc 1:179–185
2. Visa N, Jordan-Pla A (2018) ChIP and ChIP-related techniques: expanding the fields of application and improving ChIP performance. Methods Mol Biol 1689:1–7
3. Nelson JD, Denisenko O, Sova P et al (2006) Fast chromatin immunoprecipitation assay. Nucleic Acids Res 34:e2
4. Small EC, Maryanski DN, Rodriguez KL et al (2021) Chromatin immunoprecipitation (ChIP) to study DNA-protein interactions. Methods Mol Biol 2261:323–343

5. Gao H, Zhao C (2018) Analysis of protein-DNA interaction by chromatin immunoprecipitation and DNA tiling microarray (ChIP-on-chip). Methods Mol Biol 1689:43–51
6. Reimer JJ, Turck F (2010) Genome-wide mapping of protein-DNA interaction by chromatin immunoprecipitation and DNA microarray hybridization (ChIP-chip). Part A: ChIP-chip molecular methods. Methods Mol Biol 631:139–160
7. Nakato R, Sakata T (2021) Methods for ChIP-seq analysis: a practical workflow and advanced applications. Methods 187:44–53
8. Rotem A, Ram O, Shoresh N et al (2015) Single-cell ChIP-seq reveals cell subpopulations defined by chromatin state. Nat Biotechnol 33:1165–1172
9. Dai C, Chen X, Li J et al (2020) Transcription factors in ferroptotic cell death. Cancer Gene Ther 27:645–656
10. Chen X, Li J, Kang R et al (2021) Ferroptosis: machinery and regulation. Autophagy 17: 2054–2081
11. Tang D, Chen X, Kang R et al (2021) Ferroptosis: molecular mechanisms and health implications. Cell Res 31:107–125
12. Mao C, Wang X, Liu Y et al (2018) A G3BP1-interacting lncRNA promotes ferroptosis and apoptosis in cancer via nuclear sequestration of p53. Cancer Res 78:3484–3496
13. Jiang Y, Mao C, Yang R et al (2017) EGLN1/c-Myc induced lymphoid-specific helicase inhibits ferroptosis through lipid metabolic gene expression changes. Theranostics 7:3293–3305
14. Wang M, Mao C, Ouyang L et al (2019) Long noncoding RNA LINC00336 inhibits ferroptosis in lung cancer by functioning as a competing endogenous RNA. Cell Death Differ 26:2329–2343
15. Chu C, Chang HY (2016) Understanding RNA-chromatin interactions using chromatin isolation by RNA purification (ChIRP). Methods Mol Biol 1480:115–123
16. Chu C, Qu K, Zhong FL et al (2011) Genomic maps of long noncoding RNA occupancy reveal principles of RNA-chromatin interactions. Mol Cell 44:667–678
17. Chu C, Quinn J, Chang HY. (2012) Chromatin isolation by RNA purification (ChIRP). J Vis Exp
18. Huang G, Xiang Z, Wu H et al (2022) The lncRNA BDNF-AS/WDR5/FBXW7 axis mediates ferroptosis in gastric cancer peritoneal metastasis by regulating VDAC3 ubiquitination. Int J Biol Sci 18:1415–1433
19. Wang Z, Chen X, Liu N et al (2021) A nuclear long non-coding RNA LINC00618 accelerates ferroptosis in a manner dependent upon apoptosis. Mol Ther 29:263–274
20. Wu Y, Zhang S, Gong X et al (2020) The epigenetic regulators and metabolic changes in ferroptosis-associated cancer progression. Mol Cancer 19:39
21. Huang J, Wang J, He H et al (2021) Close interactions between lncRNAs, lipid metabolism and ferroptosis in cancer. Int J Biol Sci 17:4493–4513

Chapter 4

PAR-CLIP Assay in Ferroptosis

Xiangfei Xue, Manyuan Wang, Xiao Zhang, Lifang Ma, and Jiayi Wang

Abstract

Ferroptosis is a regulatory cell death process that is accompanied by large amounts of iron ion accumulation and lipid peroxidation. Photoactivated ribonucleoside-enhanced crosslinking and immunoprecipitation (PAR-CLIP) is a method used to identify the binding sites of RNA-binding proteins (RBPs) on target RNAs with high resolution at the nucleotide level. By inserting photosensitive ribonucleoside analogs into new RNA transcripts of living cells, characteristic mutations can be generated during reverse transcription and be used to accurately locate the crosslinking position of RNAs and RBPs. The use of PAR-CLIP to detect interactions and determine precise crosslinking sites between RNAs and RBPs, or to search for RNAs upstream or downstream of ferroptosis pathways genes through known proteins, can help to clarify and verify the occurrence and regulation mechanisms of the various signaling pathways of ferroptosis. Furthermore, it may reveal new targets for ferroptosis detection and improve the treatment efficiency of ferroptosis-related diseases such as cancer and neurodegenerative diseases. Here, we introduce a specific PAR-CLIP protocol for monitoring the ferroptosis process.

Key words Ferroptosis, Photoactivatable-ribonucleoside-enhanced crosslinking and immunoprecipitation (PAR-CLIP), Crosslinking and immunoprecipitation (CLIP), RNA, RNA binding protein

1 Introduction

Ferroptosis was first discovered and reported by Dixon as a novel form of cell death in 2012, characterized by the accumulation of large amounts of iron ions and lipid peroxidation during the process [1]. There are multiple regulatory pathways involved that mediate ferroptosis. Currently, numerous studies have indicated that microRNAs (miRNAs) and long non-coding RNAs (lncRNAs) are crucial mediators of ferroptosis regulation [2]. Additionally, RNA-binding proteins (RBPs) can bind to specific RNA molecules and affect the downstream signaling pathways [3, 4]. These interactions are important for explaining the occurrence and regulation mechanisms related to ferroptosis.

Authors Xiangfei Xue, Manyuan Wang, Xiao Zhang have equally contributed to this chapter.

Guido Kroemer and Daolin Tang (eds.), *Ferroptosis: Methods and Protocols*, Methods in Molecular Biology, vol. 2712, https://doi.org/10.1007/978-1-0716-3433-2_4,

As part of post-transcriptional regulation of eukaryotic gene expression, RBPs can interact with different types of RNA molecules to control their maturation, stability, transport, and translation [5–7]. To systematically and extensively study the interactions between RBPs and various RNAs, ultraviolet (UV) crosslinking and immunoprecipitation (CLIP) and its derivative methods have been produced. CLIP was first established by Ule in 2003 [8]. Through UV crosslinking in vivo, RNA and its interacting proteins form a complex and covalently crosslink at the specific binding site. After cell lysis, the target protein-RNA complex is enriched and purified by immunoprecipitation (IP) using specific antibodies against the target protein or a fusion label on the target protein. The purified complex is de-crosslinked, then the RNA is extracted to obtain the RNA molecules that were specifically bound to the target protein. The RBP recognition sequence can be determined by reverse transcription and Sanger sequencing [9]. However, the CLIP process is long, and the library yield and reproducibility are poor. The method is also limited by low crosslinking efficiency (254 nm UV crosslinking), low crosslinked RNA content, and degradation problems. In addition, the resolution of this method is limited if multiple RNAs bind to the target protein and cannot be accurately identified [8, 10, 11].

To improve the resolution of CLIP, Hafner introduced nucleotide analogs in 2010, and invented photoactivatable ribonucleoside-enhanced crosslinking and immunoprecipitation (PAR-CLIP) [12, 13]. It is a transcription method that recognizes RBP binding sites on target RNAs with nucleotide-level resolution. This method can be easily applied to any protein that directly contacts RNA [14]. Photoactive ribonucleoside analogs (4-thiouridine (4-SU) and 6-thioguanosine (6-SG), and others) are inserted into the new RNA transcripts, which is a core step of this method. In cells exposed to 365 nm UV radiation, photoactive ribonucleoside-labeled RNA is induced to bind to RBPs [15, 16]. The target RBP is then immunoprecipitated, followed by the crosslinked and precipitated RNA being isolated. RNA is reverse transcribed into a cDNA library. Simultaneously, the characteristic mutation (T-to-C for 4-SU and G-to-A for 6-SG) is introduced during reverse transcription at the crosslinking position. Depending on the mutation position in the cDNA sequence, the crosslinking position can be accurately located [10, 17].

The advantages of PAR-CLIP are: (1) Increase the crosslinking efficiency by 100 to 1000 times [13]; (2) Use the mutation of the site to determine the RNA-binding site and obtain the real cross-linked RNA sequence; (3) With the recognition of RNA-binding sites, the resolution is improved with reduced background interference; (4) The putative binding sites (sequence clusters) can be sorted by the number of transferred nucleotides [18].

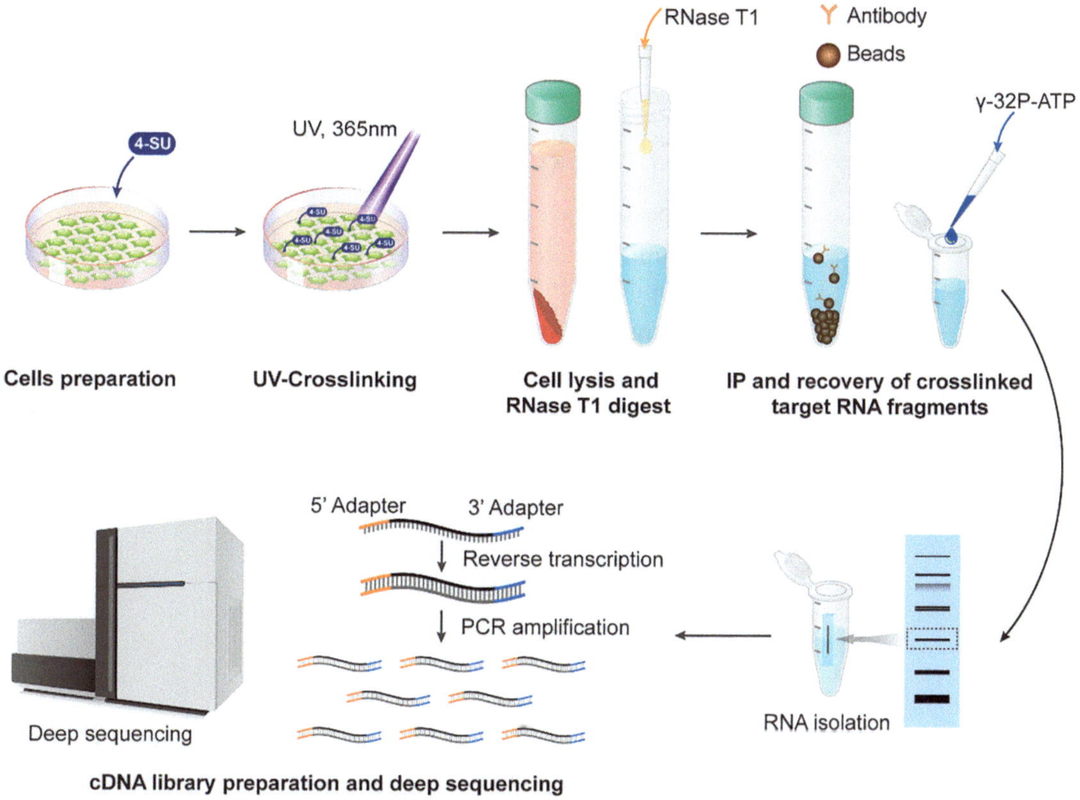

Fig. 1 Graphical protocol overview of PAR-CLIP assay methods

PAR-CLIP can be used to detect the interactions between RNAs and proteins to help verify the influence of RBPs on ferroptosis, as well as clarify the mechanisms involving ferroptosis-related signaling pathways by identifying targets of RNAs, especially miRNAs and lncRNAs [19]. Investigating ferroptosis from this perspective may provide several new ferroptotic targets and help enhance the efficacy of ferroptosis-based therapy for future treatments for diseases, such as cancer, neurodegenerative disorders, ischemia and reperfusion injury, and inflammatory diseases [20–22]. In this chapter, we introduce our PAR-CLIP methods in ferroptosis (Fig. 1).

2 Materials

The following materials will be covered by the present protocol:

1. Antibody.
2. Acid phenol.
3. Calf intestinal alkaline phosphatase.

4. Cell culture medium.
5. Cell scraper.
6. Chloroform.
7. Dephosphorylated buffer: 50 mM Tris-HCl, pH 7.9, 100 mM NaCl, 10 mM $MgCl_2$, 1 mM DTT.
8. D-Tube Dialyzer Midi Tube.
9. Formamide gel loading solution: 50 mM EDTA, 0.05% (w/v) bromophenol blue, formamide ad 100%.
10. Gel extraction kit.
11. High-speed centrifuge.
12. Isoamyl alcohol.
13. Liquid nitrogen.
14. Magnetic separation device.
15. Metal bath.
16. Non-radioactive ATP.
17. PCR instrument.
18. Phosphor imager.
19. RNase T1 (1000 U/μL).
20. Polynucleotide kinase (PNK) buffer: 50 mM Tris-HCl, pH 7.5, 50 mM NaCl, 10 mM $MgCl_2$, 5 mM DTT.
21. Primers.
22. Protease K (20 mg/mL).
23. Protease K Buffer: 100 mM Tris-HCl, pH 7.5, 150 mM NaCl, 12.5 mM EDTA, 2% (w/v) SDS.
24. Protein G magnetic beads.
25. Rotational Mixer.
26. SDS-PAGE loading buffer: 10% glycerol (v/v), 50 mM Tris-HCl, pH 6.8, 2 mM EDTA, 2% SDS (w/v), 100 mM DTT, 0.1% bromophenol blue.
27. Spectrolinker.
28. Superscript III reverse transcriptase (200 U/μL).
29. Taq polymerase (5 U/μL).
30. T4 polynucleotide kinase (10 U/μL).
31. T4 RNA ligase 1 (5 U/μL).
32. T4 RNA ligase 2 (5 U/μL).
33. Ultraviolet transmission device.
34. Water bath.
35. 1.5 mL low adhesion microcentrifuge tube.
36. 1.5 mL microcentrifuge tube.

37. 15 cm cell culture plates.
38. 15 mg/mL glycogen.
39. 15 mL centrifuge tube.
40. 100 μM 5′ adapter oligonucleotide.
41. 100 μM adenylated 3′ adapter.
42. 15% acrylamide gel.
43. 100% ethanol.
44. 1× NP40 lysis buffer: 50 mM HEPES, pH 7.5, 150 mM KCl, 2 mM EDTA, 1 mM NaF, 0.5% (v/v) NP40, 0.5 mM DTT, complete EDTA-free protease inhibitor cocktail (Roche).
45. 1× SDS running buffer: 25 mM Tris-HCl, 0.25 M glycine, 0.1% SDS (w/v).
46. 1× transfer buffer: 1 NuPAGE Transfer Buffer, 20% MeOH.
47. 1× TBE buffer: 10.8 g Tris, 0.744 g EDTA, 5.5 g boric acid.
48. 1× DNA loading dye.
49. 10× dNTPs: 2 mM dATP, 2 mM dCTP, 2 mM dGTP, 2 mM dTTP.
50. 10× RNA Ligase Buffer: 0.5 M Tris-HCl, pH 7.6, 0.1 M $MgCl_2$, 0.1 M 2-mercaptoethanol, 1 mg/mL acetylated BSA.
51. 2.5% agarose gel.
52. 3 M NaCl.
53. 4–12% polyacrylamide gel.
54. 4-SU.
55. 50 mL centrifuge tube.
56. 50% DMSO.
57. 5× first-strand synthesis buffer.
58. 6-SG.
59. 0.1 M DTT.
60. 0.4 μg/mL ethidium bromide.
61. 0.45 μm nitrocellulose membrane.
62. −20 °C refrigerator.
63. −80 °C refrigerator.
64. γ-32P-ATP.

3 Methods and Protocols

3.1 Cells Preparation

1. Expand cells in appropriate growth medium in 15 cm plates. We recommend using number of cells, which will produce 1.5–3 mL of wet cell pellets (e.g., approximately

50–200 × 10^6 H1299 cells will result from 5–20 15 cm cell culture plates), as the starting point. Grow cells to approximately 80% confluency.

2. Treat the cells with ferroptosis inducers or other stimuli for appropriate time.
3. 16 h before crosslinking, add 4-SU directly to a final concentration of 100 μM (1 M 4-SU stock solution 1:1000 v/v) (*see* **Note 1**).

3.2 UV-Crosslinking

3.2.1 UV-Crosslinking for Adherent Cells

1. Wash cells once with 10 mL ice-cold 1× PBS (pH 7.2–7.4) per plate and remove the PBS completely.
2. Place plates on a tray with ice and irradiate the cells uncovered with a dose of 0.15 J/cm^2 of 365 nm UV light in a spectrolinker equipped with 365 nm light bulbs or similar device. This will crosslink the RNA with the protein.
3. Cover the cells with 2 mL PBS per plate.
4. Scrape the cells off with a cell scraper.
5. Transfer the cell suspension to a 50 mL centrifuge tube.
6. Centrifuge at 4 °C for 5 min at 500 G-force, and discard the supernatant.
7. If IP is not performed immediately following cell lysis, please flash freeze the cell pellets in liquid nitrogen and store them at −80 °C. Cell pellets can be stored for at least 12 months at −80 °C.

3.2.2 UV-Crosslinking for Cells Grown in Suspension

1. Collect suspension cells by centrifuging them at 4 °C for 5 min at 500 G-force.
2. Resuspend cells in 10 mL PBS, then transfer the suspension to a 15 cm cell culture plate.
3. Place plates on a tray with ice and irradiate the cells uncovered with a dose of 0.2 J/cm^2 of 365 nm UV light in a spectrolinker equipped with 365 nm light bulbs or similar device. This will crosslink the RNA with the protein.
4. Transfer the cells to a 50 mL centrifuge tube.
5. Collect cells by centrifugation at 4 °C for 5 min at 500 G-force, then discard the supernatant.
6. If IP is not performed immediately following cell lysis, please flash freeze the cell pellets in liquid nitrogen and store them at −80 °C. Cell pellets can be stored for at least 12 months at −80 °C.

3.3 Cell Lysis and RNase T1 Digestion

1. Add three volumes of 1× NP40 lysis buffer to the crosslinked cell pellets in a 15 mL centrifuge tube (*see* **Note 2**).
2. Incubate and lysate the cells on ice for 10 min.

3. Centrifuge at 4 °C for 15 min at 13,000 G-force.
4. Transfer the supernatant to a new 15 mL centrifuge tube and discard the pellets.
5. Add RNase T1 (1000 U/μL) to a final concentration of 1 U/μ L (*see* **Note 3**).
6. Incubate the centrifuge tube in a water bath at 22 °C for 15 min.
7. Cool them on ice for 5 min.
8. Keep a 100 μL aliquot of cell lysate, then store that at −20 °C to control for protein expression.

3.4 Immunoprecipitation (IP) and Recovery of Crosslinked Target RNA Fragments

3.4.1 Preparation of Magnetic Beads

1. Transfer 20 μL of Protein G magnetic beads per 1 mL cell lysate (for a typical experiment, it should be approximately 120–200 μL of beads) to 1.5 mL microcentrifuge tube.
2. Clean the beads twice with 1 mL PBS.
3. Resuspend the beads in two volumes of PBS.
4. Add 0.25 mg antibody per 1 mL of the beads suspension, then incubate on a rotational mixer at room temperature for 40 min.
5. Clean the beads twice in 1 mL PBS to wash away the unbound antibody.
6. Resuspend the beads in PBS (one original bead volume).

3.4.2 IP, Second RNase T1 Digestion, and Dephosphorylation

1. Add 10 μL of freshly prepared antibody-conjugated magnetic beads per 1 mL of partial RNase T1-treated cell lysate.
2. Transfer to a 15 mL centrifugation tube and incubate for 1 h at 4 °C on a rotational mixer.
3. Collect the magnetic beads in the 15 mL centrifuge tube using the magnetic separation device.
4. Store 100 μL of the supernatant at −20 °C to control the consumption of protein. Discard the remaining supernatant.
5. Add 1 mL of 1× NP40 lysis buffer and transfer it to a 1.5 mL microcentrifuge tube (*see* **Note 4**).
6. Clean the beads twice with 1 mL of 1× NP40 lysis buffer to remove excess components.
7. Add 1× NP40 lysis buffer of one original bead volume.
8. Add RNase T1 (1000 U/μL) to a final concentration of 1 U/μ L and incubate them in a water bath at 22 °C for 15 min. Then cool on ice for 5 min.
9. Clean the beads three times in 1 mL of 1× NP40 lysis buffer. The magnetic beads are then washed twice with 400 μL dephosphorylation buffer solution.

10. Resuspend the beads in one original bead volume of dephosphorylation buffer.
11. Add calf intestinal alkaline phosphatase to a final concentration of 0.5 U/μL.
12. Incubated the centrifuge tube at 37 °C for 10 min.
13. Clean the beads twice in 1 mL of 1× NP40 lysis buffer.
14. Clean the magnetic beads twice in 1 mL of polynucleotide kinase (PNK) buffer without DTT.
15. Resuspend beads in one original bead volume of PNK buffer with DTT (*see* **Note 5**).

3.4.3 Radiolabeling of RNA Segments Crosslinked to Immunoprecipitated Proteins

1. To the bead suspension described above, add T4 polynucleotide kinase (10 U/μL) to 1 U/μL and γ-32^P-ATP to a final concentration of 0.5 μCi/μL (1.6 μM ATP) in one original bead volume.
2. Incubate the suspension for 30 min at 37 °C.
3. Add non-radioactive ATP to a final concentration of 100 μM, then incubate at 37 °C for 5 min.
4. Clean the magnetic beads five times with 800 μL of PNK buffer without DTT.
5. Resuspend the beads in 70 μL SDS-PAGE loading buffer.
6. Incubate them in a metal bath at 95 °C for 5 min to denature and release the immunoprecipitated proteins with crosslinked RNA.
7. Remove the magnetic beads from the magnetic separator and transfer the supernatant into a clean 1.5 mL microcentrifuge tube.
8. Considering that the half-life of 32^P is 14.5 days, we suggest that the sample can be stored at −20 °C for 2 weeks.

3.4.4 SDS-PAGE and Electroelution of Crosslinked RNA-Protein Complexes from Gel Slices or Electrophoresis, Transfer, and Recovery of RNA from Nitrocellulose Membrane

1. Prepare a 4–12% polyacrylamide gel.
2. Load 40 μL of the supernatant per well.
3. Run the gel at 200 V for 40 min.
4. Reserved 5 μL of the supernatant for western blot analysis to confirm the effectiveness of the IP procedure.

There are two options after this step.

Electroelution of Crosslinked RNA-Protein Complexes from Gel Slices

1. Remove the gel and install it on a glass plate. Tiny radioactive gel sheets are implanted asymmetrically (one in each of three corners of the gel sheet) to facilitate the alignment of the gel with the printed output of the phosphorus imaging paper. Wrap the gel with plastic film to avoid contamination.

2. Expose the gel to the blank phosphor imager screen for 1 h and observe it on the phosphor imager.
3. Align the gel with the top of the phosphor imager printout and use the implanted gel sheet for orientation.
4. Cut out the bands corresponding to the expected RBP size.
5. Add 800 μL of ddH_2O into the D-Tube Dialyzer Midi Tube and let it stand at room temperature for 5 min.
6. Remove the ddH_2O.
7. Transfer the removed bands to the dialyzer tube and add 800 μL of 1× SDS running buffer.
8. Electroelute the crosslinked RNA-RBP complex in 1× SDS running buffer at 100 V for 2 h.

Electrophoresis, Transfer, and Recovery of RNA from Nitrocellulose Membrane

1. Transfer proteins to a 0.45 μm nitrocellulose membrane with 1× transfer buffer at the rate of 2 mA/cm^2 current for 1 h by semi-dry blotting (*see* **Note 6**).
2. Use 1 μL of radioactive washing waste to label three corners of the nitrocellulose membrane and protein length markers of each band. Wrap the gel with plastic film to avoid contamination.
3. Expose the nitrocellulose membrane to the blank phosphor imager screen for 1 h at room temperature and observe it on the phosphor imager.
4. Print the image from the phosphor imager to the transparent film of the projector. Make sure the image is scaled to 100% for printing.
5. Align the gel with the top of the phosphor imager printout and use the implanted gel sheet for orientation.
6. Cut the bands on the nitrocellulose membrane corresponding to the expected size of the RBP of interest through the transparent membrane and the membrane directly below.
7. Further cut the nitrocellulose into five small pieces and transfer them into the 1.5 mL low adhesion tube (*see* **Note 7**).

3.4.5 Proteinase K Digestion

The next steps are slightly different based on the selection of electroelution or nitrocellulose membrane in the previous step.

Selection of Electroelution in the Previous Step

1. Add 2× Protease K Buffer solution at the same volume as the eluent.
2. Add protease K (20 mg/mL) to a final concentration of 1.2 mg/mL.
3. Incubate at 55 °C for 30 min.

Selection of Nitrocellulose Membrane in the Previous Step

1. Add 400 μL of 1× Protease K Buffer solution to the nitrocellulose membranes.
2. Add about 2 mg protease K. Vortex, centrifuge briefly.
3. Incubate at 55 °C for 1 h and 30 min (*see* **Note 8**).
4. The next steps are the same.
5. Add two volumes of acid phenol/chloroform/isoamyl alcohol (25:24:1, pH 4.0) to directly extract RNA with Protease K.
6. Vortex for 15 s and centrifuge at 4 °C for 5 min at 14,000 G-force.
7. Transfer the aqueous phase into a new 1.5 mL low adhesion microcentrifuge tube without disturbing the organic phase or interphase. If either phase is accidentally disturbed, centrifuge the sample again and repeat this step.
8. Add one volume of chloroform to the recovered aqueous phase to remove the residual phenol.
9. Vortex for 15 s and centrifuge at 4 °C for 5 min at 14,000 G-force.
10. Transfer the aqueous phase into a new 1.5 mL low adhesion microcentrifuge tube without disturbing the organic phase or interphase.
11. Add 1/10 volume of 3 M NaCl, 1 μL 15 mg/mL glycogen and three volumes of 100% ethanol.
12. Invert the tube for at least five times and incubate it at −20 °C or − 80 °C for 20 min.
13. If kept at −80 °C, RNA can be safely stored as ethanol precipitation for several months.

3.5 cDNA Library Preparation and Deep Sequencing

3.5.1 3′ Adapter Ligation

1. Prepare the following reaction mixture for connecting the 3′ adenylated adapter. Multiply the volume by the number of connecting reactions to be performed, plus the additional volume considering the pipet error: 2 μL of 10× RNA Ligase Buffer without ATP, 6 μL 50% DMSO, and 1 μL 100 μM adenylated 3′ adapter.
2. Add 9 μL of the reaction mixture to each sample.
3. Prepare a 40 fmol of 1:100 diluted sample of 5′ – 32^{P} labeled RNA size marker (*see* **Note 9**).
4. Incubate the tube at 90 °C for 1 min to denature the RNA. Immediately place the tube on ice for 2 min.
5. Add 1 μL of T4 RNA ligase 2 (5 U/μL), mix gently. Then incubate overnight on ice in a cold room.
6. Add 20 μL formamide gel loading solution and incubate the samples at 95 °C for 2 min.

7. Fill the samples in two adjacent wells of a 20-well 15% acrylamide gel. Ensure that the different samples are properly separated, usually at a distance between two wells, in order to avoid cross-contamination.
8. Marker reactions are loaded on both ends of the gel.
9. Run the gel with 1× TBE buffer at 30 W for 45 min until the bromophenol blue dye is close to the bottom of the gel.
10. Remove the gel and install it on a glass plate. Tiny radioactive gel sheets are implanted asymmetrically (one in each of three corners of the gel sheet) to facilitate the alignment of the gel with the printed output of the phosphorus imaging paper. Wrap the gel with plastic film to avoid contamination.
11. Expose the gel to the phosphor imager screen for at least 1 h. If the radioactivity of the recovered RNA is weak, the gel can be exposed overnight and the cassette tape can be stored at −20 °C.
12. Align the gel precisely with the printout according to the position of the three radioactive gel sheets.
13. Cut the 19–35 nt bands defined by the ligation products above the 19 nt and 35 nt marks.
14. Put gel tablets into 1.5 mL low adhesion microcentrifuge tubes respectively and add 0.3 M NaCl (>300 μL).
15. Eluent the ligation product into NaCl by constant stirring at 4 °C for overnight.
16. Take the supernatant and add 1 μL of Glycogen solution, mix thoroughly, and then add three volumes of 100% ethanol.
17. Invert the tube over at least five times to make it completely mixed, then incubate it at −20 °C or − 80 °C for 20 min.
18. If kept at −80 °C, RNA can be safely stored as ethanol precipitation for several months.

3.5.2 5′ Adapter Ligation

1. Dissolve the pellet in 9 μL ddH_2O.
2. Combine 1 μL of 100 μM 5′ adapter oligonucleotide, 2 μL of 10× RNA Ligase Buffer with ATP, and 6 μL of 50% aqueous DMSO to prepare the reaction mixture.
3. Add 9 μL of the mixture to the samples.
4. Incubate the tube at 90 °C for 1 min to denature the RNA. Immediately place the tube on ice for 2 min.
5. Add 2 μL of T4 RNA ligase 1 (5 U/μL), mix gently, and incubate at 37 °C for 1 h.
6. Add 20 μL formamide gel loading solution and incubate the samples at 95 °C for 2 min.

7. Fill them in two adjacent wells of a 20-well 12% acrylamide gel. Ensure that the different samples are properly separated, usually at a distance between two wells in order to avoid cross-contamination.
8. Marker reactions are loaded on both ends of the gel.
9. Run the gel with 1× TBE buffer at 30 W for 45 min until the bromophenol blue dye is close to the bottom of the gel.
10. Image the gel and remove the new ligation product.
11. Put gel tablets into 1.5 mL low adhesion microcentrifuge tubes respectively and add 0.3 M NaCl (> 300 μL).
12. Eluent the ligation product into NaCl by constant stirring overnight at 4 °C.
13. Take the supernatant and add 1 μL of Glycogen solution, mix well, and add three volumes of 100% ethanol.
14. Invert the tube over at least five times to make it completely mixed, then incubate it at −20 °C or − 80 °C for 20 min.
15. If kept at −80 °C, RNA can be safely stored as ethanol precipitation for several months.

3.5.3 Reverse Transcription

1. Dissolve the pellet in 5.6 μL ddH_2O.
2. Incubate the tube at 90 °C for 30 seconds to denature the RNA, then transfer the tube to an incubator at 50 °C.
3. Combine 1.5 μL 0.1 M DTT, 3 μL 5× first-strand synthesis buffer, and 4.2 μL 10× dNTPs to prepare the reaction mixture.
4. Add 8.7 μL of the reaction mix to each sample and incubate at 50 °C for 3 min.
5. Add 0.75 μL Superscript III reverse transcriptase (200 U/μL) and incubate at 50 °C for 2 h.
6. Add 85 μL ddH_2O and mix thoroughly.

3.5.4 PCR Amplification

1. Combine 40 μL 10× PCR buffer, 40 μL 10× dNTPs, 2 μL 100 μM 5′ primer, 2 μL 100 μM 3′ primer, and 272 μL ddH_2O to prepare the mixture.
2. Use Taq polymerase (5 U/μL) to perform a standard 100 μL PCR.
3. 89 μL of the mixture should be used for the pilot PCR reaction, while the remainder is used for the large-scale PCR reaction.
4. Add 10 μL of the cDNA solution and 1 μL of the Taq polymerase (5 U/μL) to 89 μL reaction mixture.
5. Use the following cycling conditions: 94 °C for 45 s, 50 °C for 85 seconds, 72 °C for 60 s.

6. To determine the number of cycles required to amplify the cDNA library, remove 10 μL aliquots every three cycles from the 12th cycle to the 30th cycle.
7. Analyze the samples on a 2.5% agarose gel. The PCR product may appear as a double band. The expected length of the high band is about 95–110 nt, and the expected length of the low band is 65 nt, corresponding to the direct connection product of the 3′ adapter and the 5′ adapter.
8. Define the optimal number of cycles for cDNA amplification. This should occur during the PCR exponential amplification phase, about five cycles away from the PCR amplification saturation level.
9. Perform the PCR three times with 100 μL reaction volume using the optimal cycle number, then reanalyze the product on a 2.5% agarose gel.
10. Add three volumes of 100% ethanol to the mixture of above reactions to precipitate and concentrate the DNA.
11. Add 60 μL 1× DNA loading dye.
12. Run the sample in two wells of a 2.5% agarose gel and use 0.4 μg/mL ethidium bromide to remove any amplified 5′ adapter-3′ adapter products, which would affect the sequencing results if not removed.
13. Observe the DNA on a UV transmission device and cut out the corresponding 85–110 nt band with a clean scalpel.
14. Use a gel extraction kit to purify the DNA according to the manufacturer's instructions.
15. Recover the cDNA in 30 μL elution buffer.
16. Submit 10 μL of the purified cDNA to Illumina for sequencing.

3.6 Bioinformatic Analysis

The purpose of this biological analysis is to identify the binding sites of RNA and RBPs and provide quality control for experimental results [23].

Hundreds of millions of sequence reads can be obtained for each sample using Illumina sequencing. Next, use the biological analysis pipeline to analyze the sequence reads and compare them with the reference human genome. This allows up to one error (replacement, insertion, or deletion) to capture reads with crosslinking-induced mutation. Divide the overlapping sequence reads into groups and map them to the transcriptome. These overlapping sequence reads can be annotated and classified to obtain meaningful insights into the protein binding sites, such as sites located in exons or introns, coding sequences, or untranslated sequences [14].

Then, the frequency and distribution of the mutations introduced by photoactive ribonucleoside analogs can be calculated to rank the above groups. This can help to find the group with the strongest interaction [24, 25]. Furthermore, additional experiments are needed to link RNA binding to the regulation of targeted mRNAs or to the phenotype caused by RBPs knockout, overexpression, or mutation [26].

4 Notes

1. 100 μM 6-SG can also be used to act as a light-activated ribonucleoside. The crosslinking efficiency of 6-SG is lower than that of 4-SU [27].
2. Use 2% NP40 lysis buffer for 30 min during cell lysis to replace the original protocol, which can increase the lysis of nuclear membrane and the recovery rate of RBPs in the lysate [28].
3. Other RNA enzymes can be used in place of RNase T1 during RNA digestion, thereby altering the length distribution of RNA fragments and reducing sequence bias caused by various nucleases [29].
4. Make sure not to exceed the maximum salt concentration for the antibody to recognize its antigen.
5. Long-term exposure to high concentration of DTT will destroy the magnetic beads.
6. The advantage of using nitrocellulose membrane to recover RNA-RBP complexes from gel is that it can remove non-specific immunoprecipitated RNA and strictly purify crosslinked RNA-RBP complexes [30].
7. All manipulations of the small RNAs use low adhesion siliconized tubes. The minute amounts of small RNAs recovered after gel purification will be easily adsorbed on the wall of a standard tube.
8. Change the temperature of protease K digestion to 50 °C for 30 min, which can reduce the self-digestion of protease K, thus increasing the activity of protease on immunoprecipitated proteins [28].
9. For the standard PAR-CLIP experiment, we used a mixture of 19 nt and 35 nt size markers. This controls ligation successfully and indicates the length of the band that needs to be cut from the gel.

References

1. Dixon SJ, Lemberg KM, Lamprecht MR et al (2012) Ferroptosis: an iron-dependent form of nonapoptotic cell death. Cell 149(5): 1060–1072
2. Chen X, Li J, Kang R, et al (2021) Ferroptosis: machinery and regulation. Autophagy 17(9): 2054–2081
3. Gerstberger S, Hafner M, Tuschl T (2014) A census of human RNA-binding proteins. Nat Rev Genet 15(12):829–845
4. Castello A, Hentze MW, Preiss T (2015) Metabolic enzymes enjoying new partnerships as RNA-binding proteins. Trends Endocrinol Metab 26(12):746–757
5. Moore MJ, Proudfoot NJ (2009) Pre-mRNA processing reaches back to transcription and ahead to translation. Cell 136(4):688–700
6. Sonenberg N, Hinnebusch AG (2009) Regulation of translation initiation in eukaryotes: mechanisms and biological targets. Cell 136(4):731–745
7. Martin KC, Ephrussi A (2009) mRNA localization: gene expression in the spatial dimension. Cell 136(4):719–730
8. Ule J, Jensen KB, Ruggiu M et al (2003) CLIP identifies Nova-regulated RNA networks in the brain. Science 302(5648):1212–1215
9. Ule J, Jensen K, Mele A et al (2005) CLIP: a method for identifying protein-RNA interaction sites in living cells. Methods 37(4): 376–386
10. Konig J, Zarnack K, Luscombe NM et al (2012) Protein-RNA interactions: new genomic technologies and perspectives. Nat Rev Genet 13(2):77–83
11. Guil S, Soler M, Portela A et al (2012) Intronic RNAs mediate EZH2 regulation of epigenetic targets. Nat Struct Mol Biol 19(7):664–670
12. Hafner M, Landthaler M, Burger L et al (2010) PAR-CliP--a method to identify transcriptome-wide the binding sites of RNA binding proteins. J Vis Exp 41:e2034
13. Hafner M, Landthaler M, Burger L et al (2010) Transcriptome-wide identification of RNA-binding protein and microRNA target sites by PAR-CLIP. Cell 141(1):129–141
14. Danan C, Manickavel S, Hafner M (2022) PAR-CLIP: a method for transcriptome-wide identification of RNA binding protein interaction sites. Methods Mol Biol 2404:167–188
15. Wagenmakers AJ, Reinders RJ, van Venrooij WJ (1980) Cross-linking of mRNA to proteins by irradiation of intact cells with ultraviolet light. Eur J Biochem 112(2):323–330
16. Blackwood EM, Kadonaga JT (1998) Going the distance: a current view of enhancer action. Science 281(5373):60–63
17. Szostak E, Gebauer F (2013) Translational control by 3′-UTR-binding proteins. Brief Funct Genom 12(1):58–65
18. Kishore S, Gruber AR, Jedlinski DJ et al (2013) Insights into snoRNA biogenesis and processing from PAR-CLIP of snoRNA core proteins and small RNA sequencing. Genome Biol 14(5):R45
19. Zhang X, Xu Y, Ma L et al (2022) Essential roles of exosome and circRNA_101093 on ferroptosis desensitization in lung adenocarcinoma. Cancer Commun 42(4):287–313
20. Yu H, Guo P, Xie X et al (2017) Ferroptosis, a new form of cell death, and its relationships with tumourous diseases. J Cell Mol Med 21(4):648–657
21. Tang D, Chen X, Kang R et al (2021) Ferroptosis: molecular mechanisms and health implications. Cell Res 31(2):107–125
22. Tang D, Kang R (2023) From Oxytosis to Ferroptosis: 10 Years of Research on Oxidative Cell Death. Antioxid Redox Signal 39(1–3): 162–165
23. Hafner M, Lianoglou S, Tuschl T et al (2012) Genome-wide identification of miRNA targets by PAR-CLIP. Methods 58(2):94–105
24. Corcoran DL, Georgiev S, Mukherjee N et al (2011) PARalyzer: definition of RNA binding sites from PAR-CLIP short-read sequence data. Genome Biol 12(8):R79
25. Mukherjee N, Jacobs NC, Hafner M et al (2014) Global target mRNA specification and regulation by the RNA-binding protein ZFP36. Genome Biol 15(1):R12
26. Khorshid M, Rodak C, Zavolan M (2011) CLIPZ: a database and analysis environment for experimentally determined binding sites of RNA-binding proteins. Nucleic Acids Res 39 (Database issue):D245–D252
27. Wheeler EC, Van Nostrand EL, Yeo GW (2018) Advances and challenges in the detection of transcriptome-wide protein-RNA interactions. Wiley Interdiscip Rev RNA 9(1): e1436
28. Garzia A, Meyer C, Morozov P et al (2017) Optimization of PAR-CLIP for transcriptome-wide identification of binding sites of RNA-binding proteins. Methods 118–119: 24–40
29. Kishore S, Jaskiewicz L, Burger L et al (2011) A quantitative analysis of CLIP methods for identifying binding sites of RNA-binding proteins. Nat Methods 8(7):559–564
30. Wang Z, Kayikci M, Briese M et al (2010) iCLIP predicts the dual splicing effects of TIA-RNA interactions. Plos Biol 8(10): e1000530

Chapter 5

Organoids Models of Pancreatic Duct Adenocarcinoma

Chunhua Yu, Rui Kang, and Daolin Tang

Abstract

Three-dimensional (3D) organoid culture is a laboratory technique used to grow and study miniature organs that mimic the structure and function of real organs in the human body. Organoids are created from stem cells or tissue samples and are grown in a 3D matrix that allows them to self-organize into a complex, three-dimensional structure. Organoids are valuable tools for studying human biology and disease, including cancer. Pancreatic ductal adenocarcinoma (PDAC) still has the worst survival rate of common malignancies, despite recent advances in cancer treatment. Preclinical studies have shown that impaired cell death pathways, including apoptosis, necroptosis, ferroptosis, pyroptosis, and alkaliptosis, promote PDAC development. Organoid models are now widely used in the study of pancreatic cancer biology, including cell death machinery. This chapter provides step-by-step protocols for generating human or mice PDAC organoids in a 3D Matrigel system.

Key words Organoids, Pancreatic cancer, Cell death, Therapy

1 Introduction

Pancreatic cancer is known to be one of the most aggressive and deadliest cancers, with pancreatic ductal adenocarcinoma (PDAC) accounting for 93% of cases, and pancreatic neuroendocrine tumors accounting for only 7%. Despite significant advancements in systemic chemotherapy, the mean 5-year survival rate for PDAC remains below 10% as of 2022. This is mainly due to several factors such as late diagnosis, early metastasis, and resistance to conventional treatments. The dysregulated cell death pathway is a hallmark of PDAC [1]. Recent studies have highlighted the dual role of ferroptosis, an iron-dependent form of cell death, in PDAC [2, 3]. On one hand, ferroptosis-induced inflammatory damage can promote PDAC development [4, 5]. On the other hand, the robust induction of ferroptosis by small molecular compounds can kill cancer cells [6–11]. To further develop innovative antineoplas-

Guido Kroemer and Daolin Tang (eds.), *Ferroptosis: Methods and Protocols*, Methods in Molecular Biology, vol. 2712,
https://doi.org/10.1007/978-1-0716-3433-2_5,

tic agents, it's essential to have a deep understanding of pancreatic tumor pathogenesis, and to validate them in clinically relevant tumor models.

Unlike traditional 2D cell culture, the new 3D cell culture system provides a more physiologically relevant environment for cells to grow and interact with their surroundings. This is achieved by embedding cells in a cell-extracellular matrix (ECM), such as Matrigel, which prevents them from attaching to the bottom of the plate and allows them to form structures that closely resemble their natural tissue counterparts. These structures, called organoids, are complex clusters of organ-specific cells derived from primary embryonic or adult tissues [12–14]. Organoids are made using pluripotent stem cells (PSCs), induced pluripotent stem cells (iPSCs), embryonic stem cells (ESCs), or progenitor cells, which are formed into 3D tissue-like structures when given a scaffolding ECM such as Matrigel or collagen. Organoids have the capacity for self-renewal, self-organization, and functionality, characterized by cell-cell interactions and polarized structures. They can be maintained through indefinite passages and preserve genetic stability [15, 16].

The organoid model provides a valuable tool for studying PDAC and testing potential therapeutics [17–20]. In this chapter, we provide detailed methods for generating 3D organoid models in the field of PDAC research.

2 Materials

2.1 Required Material or Equipment (see Note 1)

1. Dissecting scissors.
2. Curved forceps.
3. Pipettes (P2, P10, P20, P100, P200, P1000).
4. Pipet-aid.
5. Pipette tips (2 μL, 10 μL, 100 μL, 200 μL, 1000 μL).
6. 5 mL, 10 mL serological pipets.
7. Baked 9″ sterile pasteur pipets.
8. Aspirator.
9. Gloves.
10. Ice bucket.
11. 1.5 mL cryogenic tubes.
12. EVOS microscope with imaging system.
13. Centrifuge.
14. Freezing container.
15. −20 °C, −80 °C freezers, 4 °C refrigerator.
16. Cryogenic locator.

17. BCL2 level biosafety cabinet.
18. 37 °C Cell culture incubator.
19. 37 °C water bath.
20. Orbital shaker.
21. Specimen mold (size: 10 mm × 10 mm × 5 mm, Tissue-Tek Cryomold).
22. 60 mm,100 mm,150 mm tissue culture dishes.
23. 150 mm Petri dish.
24. 15 mL, 50 mL conical tube.
25. 1.5 mL microcentrifuge tube.
26. 48-well tissue culture plates.
27. 500 mL 0.2 μm filter.
28. DMEM with high glucose (Hyclone).
29. DMEM with high glucose and L-glutaminutese without sodium pyruvate (Hyclone).
30. G418 (Geneticin, 50 mg/mL, Gibco).
31. Zeocin (100 mg/mL, Invitrogen).
32. Advanced DMEM/F-12 (ThermoFisher).
33. Fetal bovine serum (FBS, Gibco).
34. 30%BSA (Sigma).
35. Primocin (500 mg/mL, Invivogen).
36. Penicillin/Streptomycin (100×, Gibco).
37. 0.25%Typsin-EDTA(Gibco).
38. PBS (Gibco).
39. GlutaMAX Supplement (100×, Gibco).
40. Growth factor reduced matrigel (Corning).
41. DNase I (Sigma): stock concentration,10 mg/mL in PBS.
42. Collagenase Crude Type XI (1 g, Sigma).
43. Y-27632 (Rock inhibitor, Sigma): stock concentration,10 mM in distilled water.
44. HEPES (1 M, pH 7.2–7.5, ThermoFisher).
45. A 83–01 (TOCRIS/ R&D systems): stock concentration at 5 mM in DMSO.
46. Human epidermal growth factor (hEGF, Peprotech): stock concentration,100 μg/mL with 0.1%BSA in PBS.
47. Human fibroblast growth factor-10 (hFGF-10, Peprotech): stock concentration at 100 μg/mL with 0.1% BSA in PBS.
48. hGastrin I (TOCRIS/ R&D systems): stock concentration, 10 μM in PBS.

49. mNoggin (Peprotech): stock concentration, 100 μg/mL with 0.1% BSA in distilled water.
50. mEGF (ThermoFisher): stock concentration, 100 μg/mL with 0.1% BSA in PBS.
51. Dispase II (ThermoFisher): stock concentration, 10 mg/mL in PBS.
52. N-acetylcysteine (Sigma): stock concentration, 500 mM in distilled water.
53. Nicotinamide (Sigma): stock concentration, 1 M in PBS.
54. B27 Supplement (50×, ThermoFisher).
55. Prostaglandin E2 (PGE2, TOCRIS/ R&D systems): stock concentration, 2 mM in DMSO.
56. HistoGel (Specimen Processing Gel, ThermoFisher).
57. Histology-Cassettes (ThermoFisher).
58. 4% PFA (Paraformaldehyde, 4% in PBS, ThermoFisher).
59. Cell recovery solution (Gibco).
60. Recovery cell culture freezing medium (Gibco).
61. Dimethyl Sulfoxide (DMSO, ThermoFisher).
62. Ethanol (ETOH 100%).
63. Distilled water.
64. 10× cell lysis buffer (Cell Signaling Technology).
65. Phenylmethylsulfonyl fluoride (PMSF,10 mM reconstituted in 100% ETOH, Cell Signaling Technology).
66. Protease Inhibitor Cocktail (100×, ThermoFisher).
67. BCA protein assay kit (Thermofisher).
68. E.Z.N.A.HP Total RNA Kit (Omega).
69. L-Wnt3a cell (ATCC).
70. 293 T-HA-Rspondin1-Fc cells (R&D systems).

2.2 Buffers and Reagents to Prepare Before Starting Isolation (see Note 2)

1. L-Wnt 3a growth medium: DMEM with high glucose containing 10% FBS, 0.4 mg G-418/mL.
2. L-Wnt 3a collecting medium: advanced DMEM/F12 containing 10% FBS, 1 × GlutaMAX.
3. R-spondin growth medium: DMEM containing 10% FBS with or without 300 μg/mL Zeocin.
4. R-spondin collecting medium: advanced DMEM/F12 containing 1× GlutaMAX and 10% FBS.
5. Human wash buffer: advanced DMEM/F-12 containing 10 mM HEPES, 1× GlutaMAX Supplement, 100 μg/mL primocin, 0.1% BSA.

6. Human organoids growth medium (HOGM): human wash buffer adding 50% Wnt3a-conditioned medium, 10% R-spondin1-Conditioned Medium, 1× B27 Supplement, 10 mM nicotinamide, 2 mM N-acetylcysteine, 50 ng/mL mNoggin, 50 ng/mL hEGF, 50 ng/mL hFGF, 10 nM hGastrin I, 500 nM A 83–01(add fresh before use), 10 μM Y-27632 (add fresh before use), 1 μM PGE2, 100 μg/mL primocin.
7. Human digestion buffer: Part I + Part II.

   Part I: advanced DMEM/F-12 containing 50% Wnt3a-conditioned medium, 10% R-spondin1-conditioned medium adding 5 mg/mL collagenase XI, 10 mM HEPES, 1× GlutaMAX supplement, 100 μg/mL primocin. This is the human digestion buffer Part I, aliquots in 15 mL conical tubes and stored at 20 °C, pre-warmed it in 37 °C water bath 30 min before starting isolation (*see* **Note 3**).

   Part II: add 10 μM Y-27632 and 10 μg/mL DNase I to Part I medium (add fresh before use).
8. Mouse wash buffer: DMEM with high glucose and L-glutaminutese, without sodium pyruvate adding 1× penicillin/streptomycin and 1% FBS.
9. Mouse digestive buffer: mouse wash buffer adding 0.125 mg/mL collagenase crude type XI and 0.125 mg/mL dispase II, aliquots in 15 mL conical tubes and stored at 20 °C, pre-warmed it in 37 °C water bath 30 min before starting isolation (*see* **Note 3**).
10. Mouse splitting buffer: advanced DMEM/F-12 adding 10 mM HEPES, 1× penicillin/streptomycin, 1× GlutaMAX supplement.
11. Mouse organoids growth medium: mouse splitting buffer adding 10% R-spondin1-conditioned medium, 10 μM Y-27632 (Rock Inhibitor), 0.5 μM A 83–01, 50 ng/mL mEGF, 50 ng/mL hFGF-10, 10 nM hGastrin I, mNoggin (50 ng/mL in final), 2 mM N-acetylcysteine, 10 mM nicotinamide, 1 × B27 supplement.

3 Methods (*see* Note 4)

3.1 Conditioned Medium Preparation (see Note 5)

3.1.1 Generating Wnt3a Conditioned Medium

1. Thaw one vial of frozen L-Wnt 3a cells in a 37 °C water bath for 1–2 min.
2. Remove the vial from the water bath and decontaminate by spraying it with 70% ETOH.
3. Transfer cells into a 100 mm dish containing pre-warmed 10 mL of L-Wnt3a growth medium and place it in a 37 °C incubator overnight.

4. Discard the medium to remove DMSO in the frozen cells. Add 10 mL fresh L-Wnt3a growth medium to the dish.
5. Change the growth medium every 2–3 days until the cells reach 80%–90% confluency. Rinse cells with pre-warmed 1×PBS (10–15 mL). Detach the cells with 2 mL trypsin-EDTA in a 37 °C incubator for 3–5 min.
6. Add 8 mL growth medium to the dish and transfer it to a 15 mL conical tube.
7. Spin the tube at 300 g for 5 min and discard the supernatant, keeping the pellet.
8. Split cells 1:5 by adding 25 mL growth medium in a 150 mm dish (this dish will be preserved for freezing cells later) and 25 mL collecting medium in 4 × 150 mm dishes.
9. For the 4 × 150 mm dishes, harvest medium after 4–5 days of culture. Keep the medium at 4 °C. This is the first batch of medium.
10. Add 25 mL fresh collecting medium to the 4 × 150 mm dishes and culture for another 3 days, harvest the medium, and this is the second batch of medium.
11. Pool the first and second batch of medium together, centrifuge at 1200 g, 4 °C for 10 min, and sterilize with a 0.2 μm filter. Discard the pellet. This is the Wnt3a conditioned medium.
12. Aliquot the Wnt3a conditioned medium into 50 mL or 15 mL conical tubes and store them at −80 °C.

3.1.2 Generating R-Spondin Conditioned Medium

1. Thaw one vial of frozen 293 T-HA-Rspondin1 cells in a 37 °C water bath for 1–2 min.
2. Remove the vial from the water bath and decontaminate by spraying with 70% ethanol.
3. Transfer the cells into a 100 mm dish containing pre-warmed 10 mL of R-spondin growth medium containing zeocin and place it in a 37 °C incubator overnight.
4. Discard the old growth medium to remove DMSO in the frozen cells, and add 10 mL fresh growth medium containing zeocin.
5. Change the growth medium containing zeocin every 2–3 days until the cells reach 70–80% confluency. Rinse the cells with pre-warmed 1 × PBS (10–15 mL) and detach the cells with 2 mL of trypsin-EDTA in a 37 °C incubator for 3–5 min.
6. Add 8 mL growth medium to the dish and transfer it to a 15 mL conical tube.
7. Spin the tube at 300 g for 5 min and discard the supernatant, keeping the pellet.

8. Split the cells 1:5 by adding 10 mL growth medium containing zeocin in a 100 mm dish (reserved for frozen cells later) and 10 mL growth medium without zeocin in four 100 mm dishes.
9. Culture the cells until 80% confluency, harvest cells from all four 100 mm plates, rinse cells with pre-warmed 1 × PBS, detach cells by incubating with 2 mL of 0.25% trypsin-EDTA at 37 °C for 2–3 min, and add 10 mL growth medium without zeocin per dish. Centrifuge at 300 g for 5 min.
10. Discard the trypsin and medium, pool the pellet from the four 100 mm dishes with 5 mL of collecting medium, and split the cell pellet 1:10 into ten 150 mm dishes, each containing 25 mL of collecting medium.
11. Harvest the medium after 1 week of culture, pool the medium from all dishes into 50 mL conical tubes, centrifuge at 1200 g for 10 min, sterile the supernatant using a 0.2 μM filter, and discard the pellet. This is the R-spondin1 conditioned medium.
12. Aliquot the R-spondin conditioned medium into 15 mL or 50 mL conical tubes and store them at −80 °C.

3.2 Generation of Human Organoids from PDAC Tumor Tissue and Normal Pancreatic Tissue

1. Thaw the Matrigel on ice for at least 1–2 h before starting the isolation procedure, ensuring that it has completely liquefied. Keep it on ice throughout the entire procedure.
2. Warm the human digestive buffer and human organoid growth medium in a 37 °C water bath for 30 min before starting the isolation.
3. Place the human wash buffer on ice at least 30 min before starting the isolation procedure, and keep it on ice throughout the entire procedure.
4. The human PDAC tumor specimen or adjacent normal tissue specimen should be quickly transferred to a biopsy container containing 10 mL of human wash buffer (per sample) on ice during transportation. The specimen typically consists of a small aliquot from a biopsy measuring 3 × 2.5 × 0.4 cm, and transportation usually takes less than 30 min (*see* **Note 5**).
5. Transfer the PDAC tumor tissue or adjacent normal pancreatic tissue from the original biopsy container into a 60 mm cell culture dish containing 4 mL of cold human wash buffer using sterile forceps.
6. Gently discard the wash buffer using a manual pipette with a 1 mL tip, being careful not to disturb the specimen.
7. Wash the specimen once more with 4 mL of cold human wash buffer, and discard the wash buffer following the same method as in **step 6**.

8. Using sterile surgical scissors, mince the specimen into small pieces measuring 0.5-1 mm in length.
9. Transfer the minced specimen into a 15 or 50 mL sterile conical tube using forceps, add 4 mL of cold wash buffer to the tube, and move the small specimen pieces (which may have attached to the wall of the tube) down to the bottom by pipetting up-and-down with a 1 mL tip containing the wash medium.
10. Centrifuge the conical tube containing the small pieces of specimen at 250-300 g, 4 °C for 5 min, and gently discard the wash buffer in the tube following the same method as in **step 6**, being careful not to disturb the small pieces of specimen in the bottom of the tube.
11. Add 3 mL of pre-warmed digestive buffer containing 10 μg/mL DNase I and 10 μM Y-27632 (added fresh before use) to the conical tube containing the small pieces of specimen.
12. Place the tube containing the specimen and digestive buffer in a 37 °C water bath for an initial digestion of 15 min. During the digestion, gently shake the tube by hand for 15–30 s every 5 min. After digestion, triturate the digested tumor or tissue 10 times by pipetting up-and-down with a 1 mL pipette tip.
13. After digestion, let the tube sit on ice for 1–2 min, allowing the larger pieces of tissue to attach to the bottom of the tube.
14. Transfer the supernatant into a cleaned 15 or 50 mL conical tube by gently pipetting with a 1 mL tip without disturbing the fractions at the bottom of the tube. Label this tube as "#1" and keep it on ice.
15. Add cold human wash buffer to tube #1 and fill the tube to a total volume of 5 mL. Centrifuge tube #1 at 250–300 g, 4 °C for 5 min.
16. Gently discard the supernatant from tube #1 using the same method as in **step 6**, and keep the tissue pellet.
17. Add 2 mL of cold wash buffer to the tube #1 pellet and mix the pellet with wash buffer by gently pipetting up-and-down with a 1 mL tip. Place the tube #1 on ice.
18. Add 3 mL of pre-warmed human digestion buffer containing DNase I and Y-27632 to the remaining undigested tissue pieces (leftover) from the first digestion.
19. Repeat **steps 12–17** to obtain "#2." If there is sufficient undigested material to continue to "#3," repeat the procedure for a maximum of 45 min.
20. Pool all the tubes (#1, #2, #3) together and centrifuge at 250 g, 4 °C for 5 min. Carefully discard the supernatants without disturbing the pellets, using the same method as in **step 6**. It is important to remove the supernatants as much as possible from the pellet and keep the pellet slightly dry.

21. Resuspend the pellet of fractions in Matrigel on ice. The volume of Matrigel to use will depend on the size of the pellet. Usually, a tumor or normal tissue pellet around 25 μL in volume is resuspended in 500 μL of Matrigel.
22. Thoroughly mix the pellet with Matrigel by gently pipetting up-and-down with a 200 μL pipette tip to avoid generating bubbles. This step should be performed on ice, and the Matrigel domes should be kept on ice until they are loaded onto the plate.
23. Load 25 μL Matrigel domes into the center of a well of a 48-well plate at room temperature by gently pipetting down with a 100 μL pipette tip, being careful to avoid generating bubbles (*see* **Note 6**).
24. After loading the Matrigel domes, quickly transfer the plate into a 37 °C cell culture incubator until the Matrigel solidifies, which usually takes 20–30 min.
25. Add pre-warmed human organoid growth medium (350–450 μL per well) to cover the Matrigel. Pipette the medium in with a 1 mL tip through the wall of the well without disturbing the Matrigel dome (*see* **Note** 7).
26. Incubate at 37 °C for 5–10 days to allow organoid formation, changing the fresh growth medium containing Y-27632 every 2–3 days during incubation. Gently discard the old medium in the well (*see* **Note 8**) and add fresh growth medium as in **step 25**.
27. Observe and record the procedure of organoid formation using an EVOS microscope daily (Figs. 1 and 2).

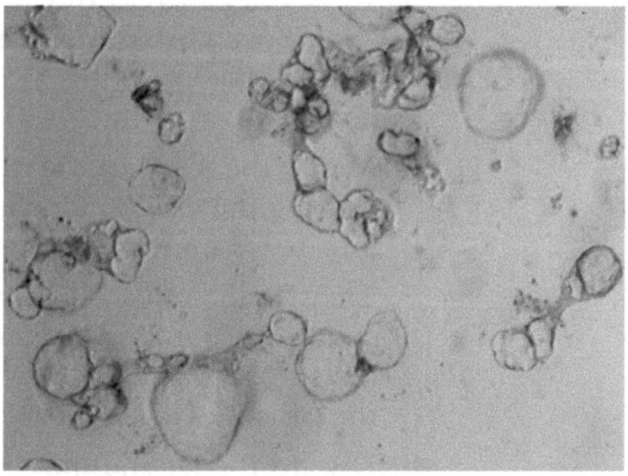

Fig. 1 Human organoids on day 6 (10× magnification) from tumor tissues of a PDAC patient donor

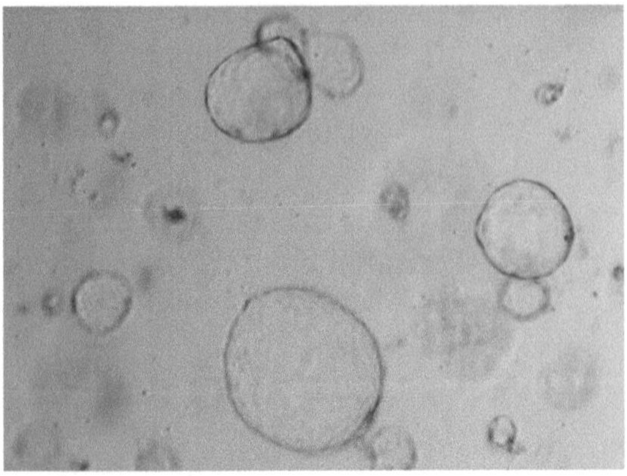

Fig. 2 Human organoids on day 6 (10× magnification) from adjacent normal tissues of a PDAC patient donor

3.3 Generation of Mouse Organoids from Normal Pancreatic Tissues of Mice

1. Thaw aliquots of 10 mg/mL DNase I and Matrigel on ice at least 1–2 h before starting the experiment.
2. Warm the mouse digestive buffer and mouse organoids growth medium in a 37 °C water bath 30 min before starting the isolation.
3. Place the mouse wash buffer on ice for at least 30 min before starting the isolation, and keep it on ice during the whole procedure.
4. Isolate the murine pancreas using sterile surgical tools (scissors and forceps). First, use 70% ethanol spray to thoroughly sterilize the mouse abdomen. Then, carefully remove the pancreas from the abdomen with sterile forceps, being careful not to rupture the intestines, and place it in a 150 mm sterile Petri dish in a cell culture hood.
5. Wash the pancreas three times with cold mouse wash buffer (5 mL per wash), and gently discard the wash buffer using a manual pipette with a 1 mL tip without disturbing the pancreas.
6. Mince the pancreas in the Petri dish with sterile scissors, keeping the dish on ice while mincing. Cut the pancreas tissue into 1–1.5 mm size pieces.
7. Transfer the minced pancreas to a sterile 15 mL conical tube containing 5 mL of cold mouse wash buffer, and let the tube sit on ice for 2–3 min to allow the fat to rise to the top and float on the surface, while the pancreas material sinks to the bottom.
8. Discard the fat and most of the mouse wash buffer by gently pipetting with a 1 mL tip without disturbing the pancreatic tissue, leaving approximately 1 mL of mouse wash buffer containing pancreatic tissue. Then, add 10 mL of mouse digestion buffer to the tube and place it in a 37 °C water bath for 20 min, gently shaking it 20–30 s by hand every 5 min.

9. Remove the tube with pancreas tissue digest from 37 °C, pipet up and down with a 10 mL pipette to break up chunks. Let the tube stand at room temperature for 2–3 min to allow the pancreas pieces to attach to the bottom by gravity.
10. Transfer 8.5 mL of supernatant to a clean Petri dish containing 10 mL of mouse wash buffer, label it wash #1, and check the Petri dish containing wash #1 under the EVOS microscope (using 4× or 10× objective) for the presence of ducts. If ducts are present, pick them (*see* **step 14** for details on how to pick ducts under the microscope).
11. Add 8.5 mL of pre-warmed mouse digestive buffer to the conical tube with the remaining pancreas and incubate at 37 °C for 10 min, gently shaking it 20–30 s by hand every 5 min.
12. After 10 min of digestion, remove the tube from 37 °C and repeat **steps 9–10** to acquire "Wash #2." Repeat to acquire "Wash #3," "Wash #4," and so on. Starting at the fourth wash (or earlier if cells start to burst and digestion looks viscous before the fourth wash), add 10 μL of 10 mg/mL DNase I to the mouse digestive buffer at each new incubation. (Fractions of wash #1 and #2, etc., if there are no visible ducts, can be preserved as a backup).
13. Continue making new washes until no more ducts are visible in the washes (usually at least 8 washes per mouse pancreas). For wild-type mice, ducts usually appear in the third-fourth wash.

    Pick the ducts: Under the 10× objective, ducts appear as small chains of cells, smaller than blood vessels, and often have lariat or cross shapes. Acinar cells usually appear as large cell masses. Use a P200 pipette with a pipette tip to gently suck up the ducts, then release the ducts into a 15 mL conical tube containing 5 mL cold mouse wash buffer and keep it on ice. Pick as many ducts as possible from the washes.
14. Centrifuge the 15 mL tubes containing the picked ducts from all the washes with 5 mL mouse wash buffer at 250–300 g for 5 min at 4 °C.
15. Discard the supernatant as much as possible without disturbing the pellets (*see* **Note 9**).

    Resuspend the pellets in Matrigel. The volume of Matrigel to use depends on the size of the pellets. Usually, a tissue pellet around 25 μL in volume should be resuspended in 500 μL Matrigel. Mix the pellet with Matrigel by gently pipetting up-and-down with a 200 μL pipette tip to avoid generating bubbles. This step should be performed on ice. Always keep the Matrigel domes on ice until loading them onto the plate.

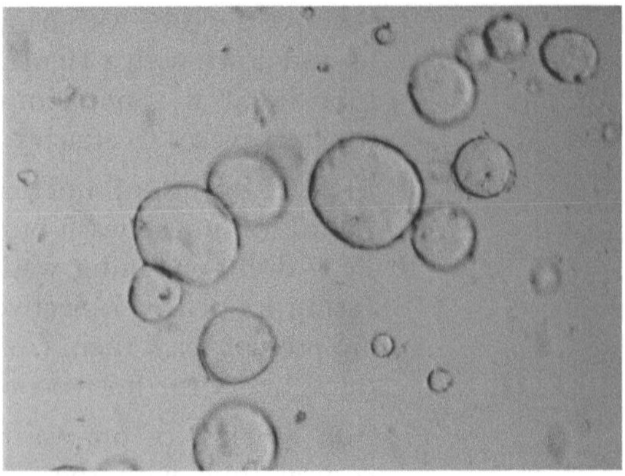

Fig. 3 Mouse organoids (10× magnification) on day 6 generated by pancreas of one mouse, WT/ C57BL/6J, Male, 12 weeks

16. Load 25 μL Matrigel domes into a well of a 48-well plate at room temperature by gently pipetting down with a 100 μL pipette tip, being careful to avoid creating bubbles (*see* **Note 10**).

    Incubate the Matrigel-ducts plate at 37 °C in a cell culture incubator for 20–30 min to allow it to solidify in the well.
17. Add mouse organoids growth medium with Rock inhibitor (Y-27632) (0.4 mL/per well) to cover the Matrigel by slowly dropping it down through the wall of the well without disturbing the Matrigel dome. Incubate at 37 °C for 5–8 days to allow organoid formation.
18. Change the mouse organoids growth medium containing Y-27632 every 2–3 days during the incubation, handling it carefully (*see* **Note 8**).
19. Check the organoid formation daily under the EVOS microscope and record the progress (Fig. 3).

3.4 Isolation of Human or Mouse Organoids from Matrigel by Generation of Pellets

1. Wash the wells containing Matrigel domes twice with cold PBS (1 mL PBS/well/48-well plate) and carefully discard the PBS without disturbing the Matrigel dome. Add 350–500 μL of Cell Recovery Solution per well for digesting the Matrigel. Seal the plate with parafilm and place it on an orbital shaker in the cold room (4 °C) for 45 min.
2. Place the Matrigel plate on ice and use a pipette to pull up the beads, then transfer them into a 15 mL conical tube on ice. Add 0.5 mL cold PBS to each well and pipette up and down a few times to remove the Matrigel as much as possible.
3. Transfer the PBS and Matrigel mixture into the same 15 mL conical tube. Keep the plate and tube on ice during this step.

4. Centrifuge the conical tube at 300 g at 4 °C for 5 min.
5. Discard the supernatant, rinse with 5 mL cold PBS, and centrifuge again at 300 g at 4 °C for 5 min.
6. Discard all the supernatant to acquire the pellet of organoids.

3.5 Preserving Organoids at Low Temperatures

1. Following **step 6** in Subheading 3.3, add 500–1000 μL of 0.25% trypsin-EDTA to the organoid pellet (the volume of trypsin depends on the size of the pellet) and trypsinize for 3–4 min in a 37 °C cell culture incubator. Then pipette the solution up-and-down 20–30 times and return to the incubator for an additional 3 min. After incubation, pipette the solution up-and-down 20–30 times again to dissociate the cells.
2. Add 10 mL wash buffer to the conical tube and centrifuge at 300 g, 4 °C for 5 min.
3. Carefully remove all the supernatant to obtain the cell pellet.
4. Resuspend the pellet in organoid freezing medium: a freezing medium containing 90% FBS and 10% DMSO or recovery cell culture freezing medium. The volume of freezing medium depends on the size of the pellet, usually using 500–1000 μL freezing medium per 50–100 μL pellet. Transfer the medium-cells mixture into 1.5 mL cryogenic tubes and place them in a freezing container at −80 °C freezer overnight. Then, move to a liquid nitrogen container for permanent storage.

3.6 Passaging Organoids

The cell pellet obtained in **step 3** of the protocol (in Subheading 3.5) is usually split at a 1:4 ratio and resuspended in 15–25 μL of Matrigel per well in a 48-well plate. Place the plate back in the 37 °C incubator for 20–30 min to allow the Matrigel dome to solidify. Then, add 350–400 μL of human or mouse organoid growth medium to maintain the growth of the organoids.

3.7 Extraction of RNA and Protein from Organoids

1. Extraction of Protein.

   Wash the pellet of organoids generated by **step 6** in the 3.3 protocol with 3 mL of cold PBS. Then, add 500–1000 μL of 1× cell lysis buffer containing 1 mM PMSF and 1× protease inhibitor cocktail. The volume of lysis buffer to use depends on the size of the pellet (add 250–500 μL of 1× lysis buffer per 50–100 μL of pellet). Mix the pellet in the lysis buffer thoroughly, vortex for 1–2 min, and incubate on ice for 10 min for digestion. Then, briefly sonicate the mixture, and centrifuge at 14,000 g at 4 °C for 10 min. Discard the pellet and keep the supernatant. Take the supernatant for protein quantification using a BCA protein assay kit and store it at −80 °C.
2. Extraction of RNA: Use the pellet of organoids generated by **step 6** in the 3.3 protocol. Rinse the pellet with 3 mL of cold PBS, then add 700 μL of GTC lysis buffer from the E.Z.N.A.

HP Total RNA Kit (Omega) and extract RNA following the manufacturer's protocol. RNA concentration is determined using a NanoDrop spectrophotometer. The RNA samples can be stored at −80 °C and utilized for various downstream applications, including RNA sequencing, PCR, qPCR, RT-PCR, microarray analysis, Northern blotting, poly-A purification, and more.

3.8 Fixation and Embedding of Organoids

1. Use the pellet of organoids generated in **step 6** of the 3.3 protocol, rinse with 3 mL cold PBS and carefully discard the PBS.
2. Resuspend the pellet in 1 mL cold PBS and transfer into a 1.5 mL eppendorf tube, then centrifuge at 300 g, 4 °C for 5 min.
3. Discard the PBS, resuspend the pellet in 500–1000 μL 4% PFA and fix overnight at 4 °C.
4. Rinse the pellet with 1 mL cold PBS twice and discard the PBS thoroughly. Let the tube with the pellet sit at room temperature.
5. Pre-warm the HistoGel in a 60–70 °C water bath for a few minutes until it turns to liquid, then place a small specimen mold (size: 10 mm × 10 mm × 5 mm) on ice.
6. Add 200–300 μL of liquid HistoGel to the pellet and gently mix them without generating bubbles.
7. Quickly transfer the mixture to the mold on ice and incubate on ice for 30–60 min to solidify.
8. Pop out the embedded organoids by hand.
9. Place the embedded organoids in a histo-cassette.
10. Place the histo-cassette in 70% ethanol for processing paraffin embedding.

4 Notes

1. All stock solutions listed in the reagents list (such as hEGF, hFGF, hGastrin I, mEGF, mNoggin, Y-27632, DNase I, and PEG2) should be aliquoted and stored at −20 °C in advance (1–2 weeks before isolation). Repeated freeze-thaw cycles are not recommended as they may cause a loss of activity for the ligands.
2. Most of the buffers and media (except for part I of the human digestive buffer) in the list should be prepared in advance, 1–2 weeks before isolation, and stored at 4 °C.

3. Do not re-freeze the thawed part I of human or mouse digestive buffer, as it may greatly decrease the activity of the enzymes.
4. To avoid potential contamination, all materials including surgical instruments, cell culture dishes, tubes, pipette tips, reagents, buffers, media, and space involved in the following procedures must maintain sterile conditions and be performed in a cell culture room.
5. Wnt3a, R-Spondins, and Noggin are essential components in organoid media as they play a key role in activating the Wnt pathway and generating organoids. The production of Wnt3a and R-Spondins conditioned medium is an economical method to acquire the ligands. It is important to prepare the conditioned medium before beginning the experiment. The media is stable when stored at −80 °C for 12 months or longer, and it will work normally for 4–6 weeks after thawing at 4 °C. Repeated freeze-thaw cycles are not recommended.
6. Keeping the specimens fresh is important for successfully generating organoids. The shorter the transportation time of the specimens, the better. Our transportation usually takes less than 30 min. Be careful when handling the specimens and avoid freezing them on dry ice or any other freezing conditions during transportation and the entire isolation procedure.
7. The cell culture plates and pipette tips do not need to be pre-warmed at 37 °C or pre-chilled at the refrigerator or on ice as in other studies. For our studies, we performed this step using plates and tips from three different conditions (1. Room temperature, 2. Pre-warmed at 37 °C for at least 24 h before isolation. 3. Pre-chilled at 4 °C or on ice). The results showed that there were no differences among the three conditions. Room temperature is a simple and convenient way.
8. The Matrigel dome is soft and fragile. Adding medium directly onto the Matrigel dome may damage it.
9. When discarding the old medium in the well, gently manually pipette with a 1 mL tip without disturbing the Matrigel dome. Do not aspirate the old medium with a vacuum aspirator as it may damage the Matrigel dome.
10. If the pellet is not visible in **step 16**, keep the remaining 300–400 μL buffer in the conical tube containing mouse ducts, add 600–700 μL cold mouse wash buffer to make a total volume of ~1 mL in the tube. Mix it and transfer this mixture to a 1.5 mL sterile eppendorf tube and centrifuge at 300 g for 5 min at 4 °C once more. If the pellet is still not visible, discard the wash buffer in the 1.5 mL tube as much as possible, keeping only ~50 μL volume liquid containing the ducts in the bottom of the tube. Then, proceed with **step 17**.

Acknowledgments

Research by D.T. and R.K. was supported by grants from the National Institutes of Health (R01CA160417, R01CA229275, and R01CA211070).

References

1. Chen X, Zeh HJ, Kang R et al (2021) Cell death in pancreatic cancer: from pathogenesis to therapy. Nat Rev Gastroenterol Hepatol 18: 804–823
2. Chen X, Kang R, Kroemer G et al (2021) Targeting ferroptosis in pancreatic cancer: a double-edged sword. Trends Cancer 7:891–901
3. Liu J, Kang R, Tang D (2021) The art of war: ferroptosis and pancreatic cancer. Front Pharmacol 12:773909
4. Dai E, Han L, Liu J et al (2020) Ferroptotic damage promotes pancreatic tumorigenesis through a TMEM173/STING-dependent DNA sensor pathway. Nat Commun 11:6339
5. Dai E, Han L, Liu J et al (2020) Autophagy-dependent ferroptosis drives tumor-associated macrophage polarization via release and uptake of oncogenic KRAS protein. Autophagy 16: 2069–2083
6. Li C, Zhang Y, Liu J et al (2021) Mitochondrial DNA stress triggers autophagy-dependent ferroptotic death. Autophagy 17: 948–960
7. Liu J, Liu Y, Wang Y et al (2023) TMEM164 is a new determinant of autophagy-dependent ferroptosis. Autophagy 19:945–956
8. Liu Y, Wang Y, Liu J et al (2021) Interplay between MTOR and GPX4 signaling modulates autophagy-dependent ferroptotic cancer cell death. Cancer Gene Ther 28:55–63
9. Li C, Liu J, Hou W et al (2021) STING1 promotes ferroptosis through MFN1/2-dependent mitochondrial fusion. Front Cell Dev Biol 9:698679
10. Song X, Liu J, Kuang F et al (2021) PDK4 dictates metabolic resistance to ferroptosis by suppressing pyruvate oxidation and fatty acid synthesis. Cell Rep 34:108767
11. Zhu S, Zhang Q, Sun X et al (2017) HSPA5 regulates ferroptotic cell death in cancer cells. Cancer Res 77:2064–2077
12. Huang L, Holtzinger A, Jagan I et al (2015) Ductal pancreatic cancer modeling and drug screening using human pluripotent stem cell- and patient-derived tumor organoids. Nat Med 21:1364–1371
13. Zhang HC, Kuo CJ (2015) Personalizing pancreatic cancer organoids with hPSCs. Nat Med 21:1249–1251
14. Lai BFL, Lu RXZ, Davenport Huyer L et al (2021) A well plate-based multiplexed platform for incorporation of organoids into an organ-on-a-chip system with a perfusable vasculature. Nat Protoc 16:2158–2189
15. Patman G (2015) Pancreatic cancer: from normal to metastases--a whole gamut of pancreatic organoids. Nat Rev Gastroenterol Hepatol 12: 61
16. Baker LA, Tiriac H, Clevers H et al (2016) Modeling pancreatic cancer with organoids. Trends Cancer 2:176–190
17. Sandhya S, Hogenson TL, Fernandez-Zapico ME (2022) Patient-derived organoids, creating a new window of opportunities for pancreatic cancer patients. EMBO Mol Med 14: e15707
18. Jeong YJ, Knutsdottir H, Shojaeian F et al (2023) Morphology-guided transcriptomic analysis of human pancreatic cancer organoids reveals microenvironmental signals that enhance invasion. J Clin Invest 133:e162054
19. Dantes Z, Yen HY, Pfarr N et al (2020) Implementing cell-free DNA of pancreatic cancer patient-derived organoids for personalized oncology. JCI. Insight 5:e137809
20. Shi X, Li Y, Yuan Q et al (2022) Integrated profiling of human pancreatic cancer organoids reveals chromatin accessibility features associated with drug sensitivity. Nat Commun 13: 2169

Chapter 6

Probing Lipid Peroxidation in Ferroptosis: Emphasizing the Utilization of C11-BODIPY-Based Protocols

Zhangshuai Dai, Wanting Zhang, Liqun Zhou, and Junqi Huang

Abstract

Ferroptosis is a form of regulated cell death that relies on iron and is characterized by the accumulation of lipid peroxides, resulting in oncotic cell swelling and eventual disruption of cellular membranes. Lipid peroxidation, a hallmark of ferroptosis, refers to the oxidative deterioration of lipids that contain carbon-carbon double bonds, particularly polyunsaturated fatty acids (PUFAs). Understanding the molecular mechanisms underlying the interplay between ferroptosis and lipid peroxidation and identifying reliable techniques for assessing lipid peroxidation levels are crucial for further advancements in this field of research. Various methods have been developed to detect lipid peroxidation levels, including C11-BODIPY (BODIPY™ 581/591 C11), liperfluo, 4-hydroxynonenal (4-HNE), malondialdehyde (MDA), Click-iT LAA (linoleamide alkyne), and liquid chromatography-mass spectrometry (LC-MS)-based epilipidomics (redox lipidomics). Currently, one of the most commonly used and effective methods is the C11-BODIPY assay, which utilizes a fluorescent probe that selectively sensitizes lipid peroxidation in cell membranes. Incorporating advanced techniques such as flow cytometry and fluorescence microscopy with C11-BODIPY dye is essential for accurate assessment of lipid peroxidation levels in ferroptosis. This chapter aims to provide comprehensive experimental protocols for detecting lipid peroxidation levels indicative of ferroptosis using C11-BODIPY staining and subsequent detection via flow cytometry and fluorescence microscopy.

Key words Ferroptosis, Lipid peroxidation, C11-BODIPY, Flow cytometry, Fluorescence microscopy

1 Introduction

Ferroptosis is a type of regulated cell death that is characterized by the accumulation of lipid peroxidation. The term "ferroptosis" was first introduced in a seminal paper published in 2012 [1]. Ferroptosis is distinguished from other forms of regulated cell death, such as apoptosis, necroptosis, pyroptosis, immunogenic cell death, cuproptosis, alkaliptosis, and disulfidptosis, based on current

Authors Zhangshuai Dai and Wanting Zhang have equally contributed to this chapter.

Guido Kroemer and Daolin Tang (eds.), *Ferroptosis: Methods and Protocols*, Methods in Molecular Biology, vol. 2712, https://doi.org/10.1007/978-1-0716-3433-2_6,

understandings [2–6]. The study of ferroptosis has provided new opportunities to investigate cell death and has significant implications for various diseases. So far, an increasing body of research has demonstrated that ferroptosis has promising therapeutic potential for conditions such as cancer, ischemia-reperfusion injury, heart failure, liver fibrosis, neurodegenerative diseases, and others [3, 7–13]. The number of diseases associated with ferroptosis is constantly increasing, and ongoing research in this field is expected to uncover additional conditions.

Lipid peroxidation is a crucial aspect of ferroptosis, which is primarily iron-dependent and involves the uncontrolled oxidation of PUFA-containing phospholipids (PUFA-PLs) in various cellular membrane structures, such as the plasma membrane, endoplasmic reticulum (ER), and mitochondria [14, 15]. Lipid peroxidation is initiated by the close association of free radicals (such as ROS and RNS) and PUFA-PLs. The chain reaction of lipid peroxidation in PUFA-PLs results in the formation of various harmful carbonyl intermediates, including MDA and 4-HNE. A range of enzymes and molecules are involved in directly or indirectly regulating lipid peroxidation, such as GPX4, system xc^{-}, DHODH, FSP1/AIFM2, TP53, NRF2, MBOAT1/2, estrogen receptor, androgen receptor, and GSH, among others [16–24]. The process and regulation mechanisms of lipid peroxidation have been extensively reviewed [25, 26].

Various techniques have been developed to detect lipid peroxidation in ferroptosis at the cellular level [25]. One approach involves measuring the degradation products of lipid peroxides such as MDA and 4-HNE. The second class of methodologies involves liquid chromatography-mass spectrometry (LC-MS)-based epilipidomics (redox lipidomics), which allows for the direct identification and quantification of specific oxygenated lipids in ferroptosis [27–32]. These methods have been extensively described in the literature [33, 34]. However, although useful, they can be relatively complex and do not provide subcellular or cellular resolution. Alternatively, specific fluorescent probes that are sensitive to lipid peroxidation, such as C11-BODIPY and liperfluo, have been widely adopted for their ease of use and ability to detect lipid peroxidation in real-time with cellular resolution. Among these probes, C11-BODIPY is the most commonly utilized and will be further discussed in the following text (*see* **Note 1**).

C11-BODIPY is a lipophilic fluorescent fatty acid analog that belongs to the BODIPY (Boron-Dipyrromethene) family of dyes. It is formally known as 4,4-difluoro-5-(4-phenyl-1,3-butadienyl)-4-bora-3a,4a-diaza-s-indacene-3-undecanoic acid. BODIPYs are well-established fluorescent dyes characterized by their exceptional fluorescence quantum yields, robust and modifiable absorption in the visible region. C11-BODIPY is specifically designed to monitor the redox state of lipid membranes in living cells, and the

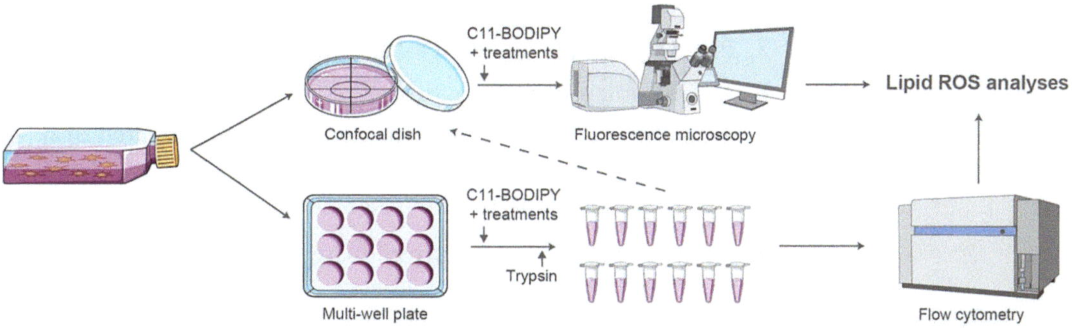

Fig. 1 Schematic diagram of lipid peroxidation assessment by C11-BODIPY in ferroptosis

fluorescence signal can be detected by flow cytometry, fluorescence microscopy, or microplate readers [35]. Upon partitioning into the lipid bilayer of cellular membranes, C11-BODIPY is susceptible to oxidation by free radicals, leading to a transition in its fluorescence emission spectrum from red to green. This spectral transition can be utilized as a quantitative method to measure lipid peroxidation in both cells and tissues [36]. Time-lapse fluorescence microscopy can be used to visualize changes in the fluorescence of oxidized C11-BODIPY in real-time [36–38]. C11-BODIPY represents a valuable technique for the surveillance of lipid peroxidation (*see* **Note 2**). However, while it has been widely used in ferroptosis research, the precise mechanisms underlying ferroptosis and the link between ferroptosis and the accumulation of lipid peroxidation require further in-depth examination (*see* **Note 8**).

The following experimental protocols outline the necessary steps and considerations for detecting lipid peroxidation levels using C11-BODIPY dye, as well as flow cytometry and fluorescence microscopy for analysis (Fig. 1). We hope that a better understanding of these detection methods will enable the scientific community to use these protocols to further develop more accurate and effective approaches for exploring the therapeutic potential of ferroptosis and lipid peroxidation (*see* **Note 20**).

2 Materials

2.1 Reagents and Consumables

1. Appropriate cell types (e.g., HT-1080).
2. Cell culture medium (e.g., DMEM, Leibovitz's L-15 medium) (*see* **Note 9**).
3. Trypsin (*see* **Note 5**).
4. C11-BODIPY dye (dissolved by DMSO, stock concentration 5 mM) (*see* **Notes 2, 4**, and **6**).
5. RSL3 and/or other ferroptosis-related chemicals.

6. DMSO (Dimethyl sulfoxide).
7. PBS (Phosphate buffered saline) (*see* **Note 13**).
8. 1.5 mL microcentrifuge tubes.
9. 4-chamber 35 mm glass bottom tissue-culture treated dish (*see* **Note 9**).
10. T25 cell culture flask.
11. Multi-well plate (e.g., 12-well plate) (*see* **Note 9**).

2.2 Equipments and Softwares

1. Flow cytometer (*see* **Note 3**).
2. Fluorescence microscopy (*see* **Note 3**).
3. Humidified carbon dioxide (CO_2) cell culture incubator.
4. Centrifuge.
5. Pipettes.
6. Vortex mixers.
7. Water baths.
8. FlowJo software.
9. ImageJ software.
10. Adobe Illustrator software.

3 Methods

3.1 Detection of Lipid Peroxidation Using Flow Cytometry

1. Seed the cells in a 12-well plate in the morning at an appropriate cell density (e.g., HT-1080, 8–10 × 10^4 per well). Optimization is required based on specific cell types. It is not recommended to process more than one 12-well plate simultaneously, as this could result in reduced processing speed and an increased variability in time intervals between samples.
2. Incubate the plate overnight in a humidified tissue culture incubator at 37 °C with 5% CO_2 (if Leibovitz's L-15 medium is used, CO_2 supplementation is typically not required).
3. The next day, replace the cell culture medium with fresh medium containing 5 μM C11-BODIPY dye and the corresponding treatments (e.g., RSL3 and/or other drugs) (*see* **Notes 4** and 7).
4. Return the plate to the tissue culture incubator. Incubate cells according to specific experimental requirements (*see* **Note 16**).
5. Collect the culture medium from each well of the 12-well plate into 12 separate 1.5 mL microcentrifuge tubes.
6. Digest cells in each well of the 12-well plate with 150 μL of trypsin for ~1 min each (please avoid over digestion of cells) (*see* **Notes 5, 10**, and **12**).

7. Stop the digestion by adding 200 μL of the corresponding medium from **step 5** into each of the 1.5 mL microcentrifuge tube.
8. Gently pipette the medium to harvest the cell suspension and transfer it to a clean 1.5 mL microcentrifuge tube.
9. Centrifuge the cells at 200–300 rcf for 3–5 min.
10. Resuspend the cells in medium containing 5 μM C11-BODIPY dye and the corresponding drugs from **step 5** (*see* **Note 11**).
11. (Optional) Place the 1.5 mL microcentrifuge tubes, which contain the cell suspensions, onto ice in a light-shielded environment if the flow cytometry instrument is not yet operational due to unforeseeable circumstances.
12. Use a flow cytometer to evaluate the level of lipid ROS using the 488 nm laser and the FL1 detector. Use a bandpass filter (e.g., 485–565 nm) for detection of oxidized C11-BODIPY signals (*see* **Note 3**).
13. Fluorescence intensity was quantified on the FL1 channel, wherein gating was employed to exclusively capture viable cells (gate was created based on the DMEM/DMSO treatment group).
14. Perform data analyses using FlowJo software. Analyze approximately 10,000 live cells in each sample.
15. Data can be exported into svg format and graph styles can be edited using Adobe Illustrator software.

3.2 Detection of Lipid Peroxidation Using Fluorescence Microscopy

1. Seed the cells in the morning at an appropriate cell density (e.g., HT-1080, 4–5 × 10^4 per well) in a 4-chamber 35 mm glass bottom tissue-culture treated dish. Fill each chamber/well with 500 μL cell culture medium. Optimization is required based on specific cell types (*see* **Notes 17** and **18**).
2. Incubate the 4-chamber cell culture dish overnight in a humidified tissue culture incubator at 37 °C with 5% CO_2 (if Leibovitz's L-15 medium is used, CO_2 supplementation is typically not required).
3. The next day night, prepare all reagents in the 1.5 mL microcentrifuge tubes, replace the cell culture medium with 500 μL fresh medium containing 5 μM C11-BODIPY dye and the corresponding treatments, such as RSL3 and/or other drugs (*see* **Note 7**). It is recommended to include negative control (DMEM/DMSO), positive control (ferroptosis inducers, e.g., RSL3), and drug treatments in the same imaging experiment to avoid misinterpretation of data.

4. During the overnight imaging process, maintain live cells in a microscopy chamber at 37 °C with 5% CO_2 (if Leibovitz's L-15 medium is used, CO_2 supplementation is typically not required).
5. Employ fluorescence microscopy to acquire fluorescence images of the C11-BODIPY dye. Implement excitation/emission of 581/591 nm (utilizing the Texas Red filter set) to distinguish the non-oxidized C11-BODIPY dye, and excitation/emission of 488/510 nm (employing the FITC filter set and incorporating a bandpass filter, such as 485–565 nm) to detect the oxidized C11-BODIPY dye. Utilize reduced laser power to minimize the potential for phototoxicity and photobleaching (*see* **Note 3**).
6. Capturing 3D images (e.g., step size 0.2–1 μm) is recommended to compare the total oxidized C11-BODIPY signals of cells (*see* **Note 14**).
7. Capturing time-lapsed 3D C11-BODIPY movies are recommended for better understanding the effects of different treatments (*see* **Notes 15, 16** and **18**).
8. Analyze the images or time-lapse movies using ImageJ or other appropriate image analysis software (*see* **Note 19**).

4 Notes

1. This chapter provides protocols for straightforward C11-BODIPY staining methods to detect lipid peroxidation in ferroptotic cells using flow cytometry and fluorescence microscopy. The experimental conditions should be optimized based on the specific cell types and research questions.
2. C11-BODIPY can be used as a probe for assays that employ either ratiometric or non-ratiometric measurement. Ratiometric assays involve quantifying two forms of the dye, C11-BODIPY^{oxidized} and C11-BODIPY$^{non\text{-}oxidized}$, and using the ratio of their fluorescent signals to calculate the level of lipid peroxidation. In non-ratiometric assays, only the oxidized form of the dye is used, and the level of fluorescence from this form is taken as a direct measure of lipid peroxidation. If the goal is to simply detect the presence of lipid peroxidation and compare the levels of peroxidation between different time points of the same sample, measuring only the oxidized C11-BODIPY signal may be sufficient. While ratiometric assays can provide better quantitative information, non-ratiometric assays can be simpler and more convenient to perform. There is currently no universally agreed standard for which method to use.

Ultimately, the choice of assay will depend on the specific experimental goals and constraints.

3. It is advisable to verify the emission wavelength range collected by the equipment (flow cytometry or fluorescence microscopy) before conducting lipid ROS measurements using C11-BODIPY. We recommend using an optical bandpass filter, which permits only a narrow range of wavelengths (i.e., 485–565 nm). Using a longpass filter that allows for a broader range of wavelengths (i.e., 450–700 nm) may lead to misleading results. The captured images could differ greatly from the observations made through the microscopy eyepiece, equipped with either an LED or mercury lamp. The source of the discrepancy is uncertain, as it may arise from the formation of dimers of BODIPY [39] or the oxidized state of C11-BODIPY, which emits light that triggers the reduced state of C11-BODIPY. As a result, a mixed light signal is generated, which does not exclusively originate from the oxidized state of C11-BODIPY.
4. There is currently no established guideline for the optimal timing of introducing the C11-BODIPY dye in ferroptosis studies. Some researchers introduce the dye simultaneously at the beginning with ferroptosis inducers, such as RSL3, while the cells are still adherent, whereas others perform C11-BODIPY staining for about 30 min after ferroptosis induction and trypsin digestion of cells. It should be emphasized that these differing methods may lead to significant variations in the assessment of lipid ROS levels. Incorporating C11-BODIPY at the beginning of ferroptotic cell death may reveal the accumulated effect of lipid peroxidation over the entire experimental period, as C11-BODIPY oxidation is irreversible [39]. In contrast, staining cells after trypsin digestion may provide evidence of the lipid peroxidation capability of cells at a specific time point.
5. If the specific experimental purpose necessitates staining C11-BODIPY after trypsin digestion, to achieve reliable results, we recommend incubating the cells with C11-BODIPY for a duration exceeding 40 min to ensure sufficient time for the oxidation of C11-BODIPY. Prolonged staining duration may enable more accurate and stable measurements of lipid peroxidation levels in the cells.
6. Upon dissolving C11-BODIPY powder, it is susceptible to oxidation/degradation even when maintained at a temperature of −80 °C for an extended duration. Therefore, to ensure its stability, it is recommended to divide it into aliquots (e.g., 20 μL per tube) for storage. When retrieved from the −80 °C

freezer, it may be stored temporarily at a temperature of −20 °C. It is advised to prevent multiple freeze-thaw cycles.

7. Given that the volume of C11-BODIPY dye utilized per well/chamber is relatively small, there is a potential for inaccuracies in the results. To address this concern, we propose the utilization of a larger centrifuge tube for mixing the C11-BODIPY dye with the medium, followed by aliquoting the mixture into smaller tubes prior to usage.
8. It is important to recognize that the lipid peroxidation capacity at a specific time point and the overall level of lipid peroxidation at a specific time point are two separate elements when studying ferroptosis. Currently, there is insufficient information regarding how the rate or magnitude of lipid peroxidation may regulate ferroptosis.
9. Various factors can affect the induction of ferroptosis and the level of lipid peroxidation. For example, the type of container used (e.g., plastic bottom versus tissue culture-treated glass bottom), contamination in cell culture, and the brand and batch number of DMEM or serum, among others. These factors should be taken into consideration when interpreting the results of experiments involving C11-BODIPY staining.
10. When using ferroptosis inducers or inhibitors that target plasma membrane proteins, caution must be exercised when conducting lipid peroxidation assays using trypsin digestion, especially if C11-BODIPY dye is added after trypsinization. Trypsin digestion can disrupt the integrity of plasma membrane proteins, leading to the loss of inducer or inhibitor efficacy and false-positive results for lipid peroxidation. To avoid this, cell scraping and subsequent flow cytometry validation or in situ observation of lipid peroxidation changes in adherent cells using fluorescence microscopy may be necessary.
11. To obtain reliable results in flow cytometry analysis after cell digestion, it is highly recommended to resuspend cells and incubate them under the original treatment conditions/drugs/dyes before detecting fluorescence signals. Discontinuing ferroptosis inducers or modulators may affect the oxidation status of C11-BODIPY before its detection, leading to alterations in experimental outcomes and interpretations.
12. To detect ferroptosis-associated lipid peroxidation, the process of trypsin digestion, centrifugation, and resuspension of cells could potentially eliminate a subset of fragile yet viable cells. As a result, it is advisable to implement fluorescence microscopy as well, for in situ detection of lipid peroxidation and cell viability.
13. In certain literatures, PBS is commonly used as a buffer for washing and resuspending cells before conducting lipid peroxidation detection via flow cytometry. Nevertheless, our results

indicate that certain formulations of PBS may provoke a time-dependent depolarization of cells, potentially promoting ferroptosis and resulting in lipid peroxidation accumulation [38]. Therefore, we recommend performing C11-BODIPY staining and detection in cell culture mediums (e.g., DMEM, Leibovitz's L-15 medium) and researchers must consider the exact composition of PBS or other salt solutions utilized in their experiments.

14. To ensure accurate quantification using confocal fluorescence microscopy, it is advisable to record whole cell signals (i.e., perform 3D imaging and sum 3D images using ImageJ software), rather than relying on a single-plane signal.
15. For time-lapse fluorescence microscopy imaging of C11-BODIPY, we suggest utilizing the 20X objective lens, as higher magnifications typically result in a reduced number of cells within the field of view, while lower magnifications may compromise image resolution.
16. Monitoring lipid peroxidation at a single time point may not provide sufficient information to fully understand the dynamics process of ferroptosis. For example, the level of lipid peroxidation is known to be time-dependent, as evidenced by HT-1080 cells exhibiting an increase in lipid peroxidation approximately 60 min after 2 μM RSL3 induction. Moreover, determining the detection time points for cells treated with ferroptosis inducers, e.g., erastin, can be difficult, as the onset of ferroptosis occurs abruptly after an extended treatment period. As such, depending exclusively on flow cytometry to assess lipid peroxidation can potentially result in inaccurate outcomes. To ensure a comprehensive understanding, it is advisable to employ time-lapse imaging of live cells to track the changes in lipid peroxidation over the entire duration of the process.
17. When acquiring time-lapse recordings of ferroptotic cells stained with C11-BODIPY using fluorescence microscopy, it is noteworthy that a variety of factors may potentially affect the homogeneity of ferroptosis induction, including variation in the contact between cells [40], plasticity of the cell state [41, 42], and meniscus issues (curved fluid-air interface) caused by surface tension [43]. Thus, to ensure the stability and accuracy of the results, it is advisable to capture images from multiple locations within the same treatment well and use a 4-chamber cell culture dish (including positive controls, such as the ferroptosis inducer RSL3, and negative controls, such as DMSO/DMEM) to validate the success of ferroptosis induction. Additionally, the 4-chamber cell culture dish represents a more practical and cost-effective option than the 8-well cell

culture dish, since the entire set of wells in the 8-well dish may not always be required.

18. When conducting time-lapse microscopy of ferroptotic cells stained with C11-BODIPY, it is important to be aware that high laser intensity could increase the proportion of the oxidized state of C11-BODIPY, leading to artificial results. To minimize potential artifacts while ensuring the quality of microscopic imaging, it is recommended to reduce the laser intensity, decrease the exposure time of the camera, or increase the time interval between consecutive time frames. After time-lapse C11-BODIPY imaging is completed, we recommend visually comparing cells in the imaged and non-imaged areas using both bright field and fluorescence channels to identify any potential artifacts that may have been introduced during the imaging process.
19. It is important to consider that when using fluorescence microscopy to visualize cells stained with C11-BODIPY, cells with low levels of lipid peroxidation may go unnoticed within the observed area if the fluorescence signal from cells possessing high levels of oxidized C11-BODIPY is excessively intense. Therefore, it is recommended to examine the images with both normal and high contrast during post-image processing.
20. To improve the accuracy of the observed outcomes for lipid peroxidation detection using C11-BODIPY or other dyes, LC-MS-based epilipidomics (redox lipidomics) analysis may be employed.

Acknowledgements

We would like to express our sincere gratitude to Ziqi Wang for her invaluable assistance in creating the schematic diagram, and to Dr. Huabin Wang for proofreading the manuscript. This work was supported by grants from the Guangzhou Basic and Applied Basic Research Foundation (grant number 202102020509), the Program for Guangdong Introducing Innovative and Entrepreneurial Teams (grant number 2017ZT07S347), and the National Natural Science Foundation of China (grant number 31701174).

References

1. Dixon SJ, Lemberg KM, Lamprecht MR et al (2012) Ferroptosis: an iron-dependent form of nonapoptotic cell death. Cell 149:1060–1072
2. Xue Q, Kang R, Klionsky DJ et al (2023) Copper metabolism in cell death and autophagy. Autophagy 16:1–21
3. Tang D, Chen X, Kang R et al (2021) Ferroptosis: molecular mechanisms and health implications. Cell Res 31:107–125
4. Tsvetkov P, Coy S, Petrova B et al (2022) Copper induces cell death by targeting

lipoylated TCA cycle proteins. Science 375: 1254–1261
5. Liu X, Nie L, Zhang Y et al (2023) Actin cytoskeleton vulnerability to disulfide stress mediates disulfidptosis. Nat Cell Biol 25: 404–414
6. Ren W, Zhao W, Cao L et al (2021) Involvement of the Actin Machinery in Programmed Cell Death. Front Cell Dev Biol 8:634849
7. Chen X, Comish PB, Tang D et al (2021) Characteristics and biomarkers of Ferroptosis. Front Cell Dev Biol 9:637162
8. Chen X, Kang R, Kroemer G et al (2021) Ferroptosis in infection, inflammation, and immunity. J Exp Med 218:e20210518
9. Conrad M, Lorenz SM, Proneth B (2021) Targeting ferroptosis: new Hope for as-yet-incurable diseases. Trends Mol Med 27:113–122
10. Jiang X, Stockwell BR, Conrad M (2021) Ferroptosis: mechanisms, biology and role in disease. Nat Rev Mol Cell Biol 22:266–282
11. Stockwell BR, Jiang X (2020) The chemistry and biology of ferroptosis. Cell Chem Biol 27: 365–375
12. Fang X, Ardehali H, Min J et al (2023) The molecular and metabolic landscape of iron and ferroptosis in cardiovascular disease. Nat Rev Cardiol 20:7–23
13. Chen X, Kang R, Kroemer G et al (2021) Targeting ferroptosis in pancreatic cancer: a double-edged sword. Trends Cancer 7:891–901
14. Zhang R, Kang R, Tang D (2023) Ferroptosis in gastrointestinal cancer: from mechanisms to implications. Cancer Lett 561:216147
15. Zou Y, Naowarojna N, Wu TW et al (2023) Dynamic regulation of ferroptosis by lipid metabolism. Antioxid Redox Signal. https://doi.org/10.1089/ars.2023.0278
16. Mao C, Liu X, Zhang Y et al (2021) DHODH-mediated ferroptosis defence is a targetable vulnerability in cancer. Nature 593:586–590
17. Doll S, Freitas FP, Shah R et al (2019) FSP1 is a glutathione-independent ferroptosis suppressor. Nature 575:693–698
18. Bersuker K, Hendricks JM, Li Z et al (2019) The CoQ oxidoreductase FSP1 acts parallel to GPX4 to inhibit ferroptosis. Nature 575:688–692
19. Jiang L, Kon N, Li T et al (2015) Ferroptosis as a p53-mediated activity during tumour suppression. Nature 520:57–62
20. Dixon SJ, Patel DN, Welsch M et al (2014) Pharmacological inhibition of cystine-glutamate exchange induces endoplasmic reticulum stress and ferroptosis. elife 3:e02523
21. Xie Y, Zhu S, Song X et al (2017) The tumor suppressor p53 limits Ferroptosis by blocking DPP4 activity. Cell Rep 20:1692–1704
22. Sun X, Ou Z, Chen R et al (2016) Activation of the p62-Keap1-NRF2 pathway protects against ferroptosis in hepatocellular carcinoma cells. Hepatology 63:173–184
23. Nakamura T, Hipp C, Santos Dias Mourão A et al (2023) Phase separation of FSP1 promotes ferroptosis. Nature. https://doi.org/10.1038/s41586-023-06255-6
24. Liang D, Feng Y, Zandkarimi F et al (2023) Ferroptosis surveillance independent of GPX4 and differentially regulated by sex hormones. Cell 186:2748–2764.e22
25. Kodali ST, Kauffman P, Kotha SR et al (2020) Oxidative Lipidomics: analysis of oxidized lipids and lipid peroxidation in biological systems with relevance to health and disease. In: Berliner LJ, Parinandi NL (eds) Measuring oxidants and oxidative stress in biological systems. Springer, Cham, pp 61–92
26. Lin Z, Liu J, Kang R et al (2021) Lipid metabolism in ferroptosis. Adv Biol (Weinh) 5: e2100396
27. Tyurina YY, Tyurin VA, Anthonymuthu T et al (2019) Redox lipidomics technology: looking for a needle in a haystack. Chem Phys Lipids 221:93–107
28. Sun WY, Tyurin VA, Mikulska-Ruminska K et al (2021) Phospholipase iPLA(2)beta averts ferroptosis by eliminating a redox lipid death signal. Nat Chem Biol 17:465–476
29. Niu J, Wan X, Yu GY et al (2022) Phospholipid peroxidation-driven modification of chondrogenic transcription factor mediates alkoxyl radicals-induced impairment of embryonic bone development. Redox Biol 56:102437
30. Mishima E, Ito J, Wu Z et al (2022) A non-canonical vitamin K cycle is a potent ferroptosis suppressor. Nature 608:778–783
31. Song X, Liu J, Kuang F et al (2021) PDK4 dictates metabolic resistance to ferroptosis by suppressing pyruvate oxidation and fatty acid synthesis. Cell Rep 34:108767
32. Lin Z, Liu J, Long F et al (2022) The lipid flippase SLC47A1 blocks metabolic vulnerability to ferroptosis. Nat Commun 13:7965
33. Martinez AM, Kim A, Yang WS (2020) Detection of ferroptosis by BODIPY 581/591 C11. Methods Mol Biol 2108:125–130
34. Li J, Kang R, Tang D (2021) Monitoring autophagy-dependent ferroptosis. Methods Cell Biol 165:163–176

35. Liu K, Liu J, Zou B et al (2022) Trypsin-mediated sensitization to Ferroptosis increases the severity of pancreatitis in mice. Cell Mol Gastroenterol Hepatol 13:483–500
36. Wang F, Graham ET, Naowarojna N et al (2022) PALP: a rapid imaging technique for stratifying ferroptosis sensitivity in normal and tumor tissues in situ. Cell Chem Biol 29:157–170.e6
37. Zou Y, Henry WS, Ricq EL et al (2020) Plasticity of ether lipids promotes ferroptosis susceptibility and evasion. Nature 585:603–608
38. Huang D, Dai Z, Wang C, et al (2023) Ferroptosis is determined by chloride ions. BioRxiv. https://doi.org/10.1101/2023.02.07.526847
39. Stockert JC (2021) Lipid peroxidation assay using BODIPY-Phenylbutadiene probes: a methodological overview. Methods Mol Biol 2202:199–214
40. Wu J, Minikes AM, Gao M et al (2019) Intercellular interaction dictates cancer cell ferroptosis via NF2-YAP signalling. Nature 572:402–406
41. Viswanathan VS, Ryan MJ, Dhruv HD et al (2017) Dependency of a therapy-resistant state of cancer cells on a lipid peroxidase pathway. Nature 547:453–457
42. Chen PH, Cai L, Huffman K et al (2019) Metabolic diversity in human non-small cell lung cancer cells. Mol Cell 76:838–851.e5
43. Barosova H, Meldrum K, Karakocak BB et al (2021) Inter-laboratory variability of A549 epithelial cells grown under submerged and air-liquid interface conditions. Toxicol In Vitro 75:105178

Chapter 7

Membrane Integrity Assay in Ferroptosis

Chao Deng and Yangchun Xie

Abstract

Ferroptosis is a form of regulated cell death that relies on the accumulation of intracellular iron and subsequent oxidative stress. Ferroptotic cell death is characterized by uncontrolled lipid peroxidation, which leads to plasma membrane damage and rupture. The loss of plasma membrane integrity results in the release of intracellular components, including damage-associated molecular patterns, and can propagate death between cells in a synchronized manner. Understanding the mechanisms of ferroptotic membrane damage is crucial to comprehending this form of cell death. This chapter provides a summary of techniques for detecting plasma membrane integrity in ferroptosis, including transmission electron microscopy analysis, flow cytometry analysis, and assessments of oxidoreductase-mediated membrane damage.

Key words Plasma membrane rupture, Ferroptosis, Transmission electron microscopy, Flow cytometry, Oxidoreductase

1 Introduction

Ferroptosis is a form of regulated cell death (RCD) that relies on intracellular iron accumulation and lipid peroxidation [1–3]. First discovered in 2012, ferroptosis is morphologically, biochemically, and genetically distinct from other types of RCD, especially apoptosis [4]. The morphological hallmark of ferroptosis is plasma membrane rupture [5, 6], which is primarily caused by radical-driven lipid oxidative membrane damage and rupture, and can be inhibited by lipophilic antioxidants (e.g., ferrostatin-1 and liproxstatin-1) [7]. Molecular compounds that target either the cystine/glutamate antiporter system xc^- or the membrane lipid repair enzyme glutathione peroxidase 4 (GPX4) trigger the accumulation of membrane lipid peroxides and ferroptosis [8–14].

Membrane phospholipids (PEs) are rich in polyunsaturated fatty acids (PUFAs) and are susceptible to ferroptotic damage [15–18]. An imbalance between oxidants and antioxidants, driven by multiple redox-active enzymes, leads to membrane peroxidation [19]. This ultimately results in the formation of plasma membrane

Guido Kroemer and Daolin Tang (eds.), *Ferroptosis: Methods and Protocols*, Methods in Molecular Biology, vol. 2712,
https://doi.org/10.1007/978-1-0716-3433-2_7,

pores a few nanometers in diameter, leading to a spike in intracellular calcium (Ca^{2+}) levels, cell permeabilization, the release of damage-associated molecular patterns (DAMPs), and death propagation between neighboring cells, which is referred to as ferroptotic rupture [5, 20, 21]. To prevent ferroptotic membrane damage and cell death, cell membrane repair systems, such as the endosomal sorting complex required for transport-III (ESCRT-III) machinery, and lipid remodeling regulators, such as the lipid flippase solute carrier family 47 member 1 (SLC47A1), are activated [18, 22–25].

Plasma membrane rupture of ferroptotic cells can trigger a range of pathophysiological processes, such as cancer cell death, acute kidney injury, pancreatitis, and Alzheimer's disease [18, 20, 26–31]. Identifying such membrane rupture can be an effective means of detecting pathological conditions associated with ferroptosis. In this chapter, we present several methods, including transmission electron microscopy (TEM) analysis, flow cytometry analysis, and assessment of oxidoreductase-mediated membrane damage, for identifying plasma membrane rupture in ferroptosis.

2 Materials

1. 2% glutaraldehyde.
2. Phosphate buffer (PBS): 10 mM $PO4^{3-}$, 137 mM NaCl, and 2.7 mM KCl pH 7.4.
3. Osmium tetroxide (OsO_4).
4. Potassium ferricyanide.
5. Ethanol.
6. Epon.
7. Uranyl acetate.
8. Lead citrate.
9. Electron microscope.
10. FACS buffer: PBS, 0.5–1% BSA, or 5–10% FBS.
11. Fluo-4 acetoxymethyl.
12. Propidium iodide (PI).
13. Flow cytometry.
14. FACSDiva software.
15. Soy phospholipid mixture.
16. Cardiolipins (Heart, Bovine).
17. 1,2-Dipalmitoyl-sn-glycero-3-phosphocholine (16: 0 PC, DPPC).
18. 1-Stearoyl-2-Arachidonoyl-sn-glycero-3-phosphoethanolamine (18:0–20:4 PE, SAPE).

19. Chloroform.
20. Nitrogen.
21. Buffer L: 20 mM HEPES, 100 mM NaCl, pH 7.4.
22. Buffer TL: 20 mM HEPES, 100 mM NaCl, 50 mM sodium citrate, 15 mM TbCl3, pH 7.4.
23. 0.1% Triton X-100.
24. Methanol.
25. 1% butylated hydroxytoluene (BHT).
26. 50 mM Dipicolinic acid (DPA).
27. 100 nM P450 reductase (POR).
28. 50 mM NADPH.
29. 120 mM Ferric chloride.
30. 10 mL Round-bottomed flasks.
31. 100 nm polycarbonate membrane.
32. 100 KD ultrafiltration tube.
33. Ultramicrotome.
34. Vacuum rotary evaporator.
35. Mini-extruder.
36. Microplate reader.
37. LC-MS/MS.

3 Methods

3.1 TEM Analysis

TEM is a high-resolution technique used to reveal structural details, size distribution, and morphology of nanoparticles made up of lipids and proteins [32]. TEM is based on the interaction between a high-energy electron beam and a thin sample, and produces an image that details the morphology, composition, and crystal structure of the specimen [32].

Studies using TEM have shown that cells undergoing ferroptosis undergo significant changes in mitochondrial ultrastructure, including a reduction in mitochondrial volume, an increase in mitochondrial membrane density, and a disappearance of mitochondrial cristae [4, 33]. Since elevated autophagy promotes ferroptosis [3, 34], autophagy-related ultrastructures such as double-membrane autophagosomes and various lysosome-related vesicles are often observed in ferroptotic cells or tissues [35]. More recently, it has been observed that ferroptotic cells exhibit cytoplasmic dilatation and that the cytoplasm, as well as the nucleus, become electron-lucent with the progression of ferroptosis. Intriguingly, the nuclei are extruded from the cytoplasm and still exist in the dilated cytosol, suggesting that nuclear membrane damage occurs prior to cytoplasmic membrane damage [36].

1. Rinse cell monolayers briefly with 0.01 M PBS.
2. Fix the cells in situ by adding 2.5% glutaraldehyde in 0.01 M PBS, pH 7.4, and incubate at room temperature for 1 h.
3. Wash the fixed monolayers three times in 0.01 M PBS buffer for 10 min each.
4. Post-fix the monolayers for 1 h at 4 °C using 1% OsO4 with 1% potassium ferricyanide.
5. Wash the monolayers again three times in 0.01 M PBS buffer for 10 min each.
6. Dehydrate the monolayers by immersing them in a graded series of alcohol (30%, 50%, 70%, and 90%) with three changes of 100% ethanol for 15 min each.
7. Infiltrate the dehydrated monolayers with epon by changing the solution three times and incubating for 1 h each time.
8. Remove the last epon change and invert beam capsules filled with resin over relevant areas of the monolayers. Polymerize at 37 °C overnight, then for 48 h at 60 °C.
9. Pop off the beam capsules and underlying cells from the bottom of the petri dish and section.
10. Cut ultrathin sections using an ultramicrotome, stain with uranyl acetate and lead citrate, and examine under an electron microscope.

3.2 Flow Cytometry Analysis

Flow cytometry is a widely used method for analyzing the expression of cell surface and intracellular molecules, characterizing and defining different cell types within a heterogeneous population, assessing the purity of isolated subpopulations, and analyzing cell size and volume [37]. Typically, fluorescence intensity produced by fluorescent-labeled antibodies detecting proteins, or ligands that bind to specific cell-associated molecules such as propidium iodide (PI) binding to DNA, is measured.

Ferroptosis, a form of cell death associated with plasma membrane damage, is characterized by the formation of membrane nanopores of a few nanometers in radius and sustained Ca^{2+} influxes prior to plasma membrane rupture [5, 20]. Flow cytometry can be used to test for irreversible plasma membrane disruption in ferroptosis by measuring intracellular Ca^{2+} levels marked with fluo-4 acetoxymethyl (Fluo-4 AM), as well as the ability of cells to exclude PI [37].

1. After treatment, collect both attached and non-attached cell populations.

2. Wash the cells with ice-cold FACS buffer (PBS, 0.5–1% BSA, or 5–10% FBS) and adjust cell concentration to 1–5 × 10^6 cells/mL.
3. Centrifuge the cells at 2500 g for 5 min at 4 °C and resuspend them in PBS containing Fluo-4 AM (1 μM).
4. Incubate the cells at 37 °C for 30 min, then wash them and resuspend them in PBS containing PI (2 μg/mL).
5. Keep the cells in the dark on ice or at 4 °C in a fridge until analysis.
6. Analyze 10,000 cells using flow cytometry (*see* **Note 1**).
7. Analyze the data using FACSDiva software (*see* **Note 2**).

3.3 Assessing Oxidoreductase-Mediated Membrane Damages

Recent research indicates that oxidoreductases, specifically NADPH-cytochrome P450 reductase (POR) and NADH-cytochrome b5 reductase (CYB5R1), play a role in phospholipid peroxidation and ferroptosis execution [38, 39]. These enzymes generate hydrogen peroxide (H_2O_2), leading to lipid peroxidation and ultimately membrane rupture [39]. Therefore, experiments that measure oxidoreductase-mediated oxidative rupture of PUFA can be utilized to investigate the damage to membranes caused by ferroptosis [40].

3.3.1 Preparation of Phospholipid Films

1. Dissolve the phospholipids (a mixture of soy phospholipids, cardiolipins, DPPC, and SAPE obtained from Avanti) in chloroform to create a 10 mg/mL stock solution at −20 °C under nitrogen protection (*see* **Notes 3** and **4**).
2. Mix 100 mL of the stock solution with 500 mL of chloroform in 10 mL round-bottomed flasks, and evaporate the solvent using a vacuum rotary evaporator with gentle and continuous rotation (150 rpm) under a water bath at 30 °C to obtain phospholipid films (*see* **Note 5**).

3.3.2 Generation of Liposomes

1. Hydrate the lipid films with 500 mL of buffer L, and subject the liposomes to 30 rounds of extrusion through a 100 nm polycarbonate membrane using a mini-extruder (*see* **Note 6**).
2. To prepare Tb^{3+} encapsulated liposomes, hydrate the lipid films with 500 mL of buffer TL.
3. Wash the liposomes 8 times with a 100 KD ultrafiltration tube by centrifuging them with buffer L for 20 min at 4 °C at a rate of 4000× *g* to remove any external Tb^{3+}, then resuspend them in 500 mL of buffer L (*see* **Note 7**).

3.3.3 Initiation of Reaction and Recording

1. Dilute 10 mL of the Tb^{3+} encapsulated liposomes (100 mg/mL) into 100 mL of buffer L supplemented with 50 mM of DPA.

2. Add 100 nM of POR, 50 mM of NADPH, and 120 mM of ferric chloride into the reaction mixture in the presence of oxygen, and measure the fluorescence signal (λex = 270/λ em = 620 nm) using a microplate reader as Ft0 (*see* **Note 8**).
3. Record the fluorescence signal at 20-s intervals approximately 120 times. After that, add 0.1% Triton X-100 to completely release the Tb^{3+}. Then measure the fluorescence signal as Ft100 (*see* **Note 9**).
4. Transfer 2 mL of the reaction buffer from different time points into 50 mL of methanol containing 1% BHT to assess the total SAPE and peroxidation of SAPE.
5. Centrifuge the samples at 20,000 g for 10 min and analyze them by LC-MS/MS.

4 Notes

1. PI fluorescence is measured in the PE-A channel, while Fluo-4 AM/Ca fluorescence is measured in the FITC channel.
2. For PI, fluorescence intensity distributions for negative (living cells) and positive (dead cells) do not overlap. Select a threshold value that yields 10% or fewer dead cells in the negative control, intermediate between the well-resolved peaks. For Fluo-4 AM, fluorescence intensity distributions for negative and positive control populations partially overlap. Choose a threshold value based on the negative control sample using the geometric triangle method, which typically yields values between 5% and 10% Fluo-4 AM/Ca positive cells in the control population [41]. We recommend analyzing samples on the same day. For extended storage (16 h) or greater flexibility in planning time on the cytometer, resuspend cells in 1–4% paraformaldehyde to prevent deterioration.
3. Because this protocol is used to test lipid peroxidation of PUFA phospholipids-mediated membrane damage, ensure that PUFA phospholipids are included in the lipid mixtures.
4. Cover the lipid carefully with nitrogen to protect it from oxygen. Even a small amount of oxygen can cause oxidative damage to the lipids over time. Optionally, argon can also be used to protect phospholipids from oxygen.
5. Gentle and continuous rotation of the flask is crucial for forming an even lipid film.
6. The extrusion process should be slow and gentle, at an approximate speed of 100 mL per second. Pre-warming the mini-extruder on a 65 °C block greatly facilitates the solubility of some phospholipids.

7. Extensively wash external Tb^{3+} ions to minimize background in the following assays.
8. In previous research [42], the fluorescence signal of Tb^{3+}/DPA chelates was detected using an emission wavelength $\lambda em = 490$ nm. However, NADPH has a strong signal under λem 490 nm but is not detected under λem 620 nm. Additionally, λem 620 nm gives a higher signal for Tb^{3+}/DPA than λem 490 nm. Therefore, λem 620 nm is chosen in this assay.
9. At each time point, liposome leakage percentage can be defined as: Leakage (t) (%) = [(Ft-Ft0)/(Ft100-Ft0)] × 100.

Acknowledgments

This work was supported by grants from the National Natural Science Foundation of China (82172656 and 81802476), and the Natural Science Foundation of Hunan Province (2021JJ40882).

References

1. Xie Y, Hou W, Song X et al (2016) Ferroptosis: process and function. Cell Death Differ 23: 369–379
2. Chen X, Kang R, Kroemer G et al (2021) Broadening horizons: the role of ferroptosis in cancer. Nat Rev Clin Oncol 18:280–296
3. Xie Y, Hou T, Liu J et al (2023) Autophagy-dependent ferroptosis as a potential treatment for glioblastoma. Front Oncol 13:1091118
4. Dixon SJ, Lemberg KM, Lamprecht MR et al (2012) Ferroptosis: an iron-dependent form of nonapoptotic cell death. Cell 149:1060–1072
5. Pedrera L, Espiritu RA, Ros U et al (2021) Ferroptotic pores induce Ca(2+) fluxes and ESCRT-III activation to modulate cell death kinetics. Cell Death Differ 28:1644–1657
6. Chen X, Comish PB, Tang D et al (2021) Characteristics and biomarkers of ferroptosis. Front Cell Dev Biol 9:637162
7. Lin Z, Liu J, Kang R et al (2021) Lipid metabolism in ferroptosis. Adv Biol (Weinh) 5: e2100396
8. Chen X, Li J, Kang R et al (2021) Ferroptosis: machinery and regulation. Autophagy 17: 2054–2081
9. Kang R, Zhu S, Zeh HJ et al (2018) BECN1 is a new driver of ferroptosis. Autophagy 14: 2173–2175
10. Liu Y, Wang Y, Liu J et al (2021) Interplay between MTOR and GPX4 signaling modulates autophagy-dependent ferroptotic cancer cell death. Cancer Gene Ther 28:55–63
11. Yang WS, SriRamaratnam R, Welsch ME et al (2014) Regulation of ferroptotic cancer cell death by GPX4. Cell 156:317–331
12. Han L, Bai L, Fang X et al (2021) SMG9 drives ferroptosis by directly inhibiting GPX4 degradation. Biochem Biophys Res Commun 567: 92–98
13. Xue Q, Yan D, Chen X et al (2023) Copper-dependent autophagic degradation of GPX4 drives ferroptosis. Autophagy 19(7): 1982–1996
14. Tong J, Li D, Meng H et al (2022) Targeting a novel inducible GPX4 alternative isoform to alleviate ferroptosis and treat metabolic-associated fatty liver disease. Acta Pharm Sin B 12:3650–3666
15. Kagan VE, Mao G, Qu F et al (2017) Oxidized arachidonic and adrenic PEs navigate cells to ferroptosis. Nat Chem Biol 13:81–90
16. Doll S, Proneth B, Tyurina YY et al (2017) ACSL4 dictates ferroptosis sensitivity by shaping cellular lipid composition. Nat Chem Biol 13:91–98
17. Yang WS, Kim KJ, Gaschler MM et al (2016) Peroxidation of polyunsaturated fatty acids by lipoxygenases drives ferroptosis. Proc Natl Acad Sci U S A 113:E4966–E4975

18. Lin Z, Liu J, Long F et al (2022) The lipid flippase SLC47A1 blocks metabolic vulnerability to ferroptosis. Nat Commun 13:7965
19. Liu J, Kang R, Tang D (2021) Signaling pathways and defense mechanisms of ferroptosis. FEBS J 289:7038
20. Riegman M, Sagie L, Galed C et al (2020) Ferroptosis occurs through an osmotic mechanism and propagates independently of cell rupture. Nat Cell Biol 22:1042–1048
21. Liu J, Zhu S, Zeng L et al (2022) DCN released from ferroptotic cells ignites AGER-dependent immune responses. Autophagy 18: 2036–2049
22. Jimenez AJ, Maiuri P, Lafaurie-Janvore J et al (2014) ESCRT machinery is required for plasma membrane repair. Science 343: 1247136
23. Dai E, Meng L, Kang R et al (2020) ESCRT-III-dependent membrane repair blocks ferroptosis. Biochem Biophys Res Commun 522: 415–421
24. Liu J, Kang R, Tang D (2021) ESCRT-III-mediated membrane repair in cell death and tumor resistance. Cancer Gene Ther 28:1–4
25. Dai E, Zhang W, Cong D et al (2020) AIFM2 blocks ferroptosis independent of ubiquinol metabolism. Biochem Biophys Res Commun 523:966–971
26. Tonnus W, Meyer C, Steinebach C et al (2021) Dysfunction of the key ferroptosis-surveilling systems hypersensitizes mice to tubular necrosis during acute kidney injury. Nat Commun 12:4402
27. Zhu D, Liang R, Liu Y et al (2022) Deferoxamine ameliorated Al(mal)(3)-induced neuronal ferroptosis in adult rats by chelating brain iron to attenuate oxidative damage. Toxicol Mech Methods 32:530–541
28. Tang D, Chen X, Kang R et al (2021) Ferroptosis: molecular mechanisms and health implications. Cell Res 31:107–125
29. Liu J, Song X, Kuang F et al (2021) NUPR1 is a critical repressor of ferroptosis. Nat Commun 12:647
30. Yang L, Ye F, Liu J et al (2022) Extracellular SQSTM1 exacerbates acute pancreatitis by activating autophagy-dependent ferroptosis. Autophagy 19(6):1733–1744
31. Liu K, Liu J, Zou B et al (2022) Trypsin-mediated sensitization to ferroptosis increases the severity of pancreatitis in mice. Cell Mol Gastroenterol Hepatol 13:483–500
32. Malatesta M (2021) Transmission electron microscopy as a powerful tool to investigate the interaction of nanoparticles with subcellular structures. Int J Mol Sci 22:12789
33. Yagoda N, von Rechenberg M, Zaganjor E et al (2007) RAS-RAF-MEK-dependent oxidative cell death involving voltage-dependent anion channels. Nature 447:864–868
34. Liu J, Kuang F, Kroemer G et al (2020) Autophagy-dependent ferroptosis: machinery and regulation. Cell Chem Biol 27:420–435
35. Friedmann Angeli JP, Schneider M, Proneth B et al (2014) Inactivation of the ferroptosis regulator Gpx4 triggers acute renal failure in mice. Nat Cell Biol 16:1180–1191
36. Miyake S, Murai S, Kakuta S et al (2020) Identification of the hallmarks of necroptosis and ferroptosis by transmission electron microscopy. Biochem Biophys Res Commun 527: 839–844
37. Holmes K, Lantz LM, Fowlkes BJ, et al (2001) Preparation of cells and reagents for flow cytometry. Curr Protoc Immunol Chapter 5: Unit 5 3
38. Zou Y, Li H, Graham ET et al (2020) Cytochrome P450 oxidoreductase contributes to phospholipid peroxidation in ferroptosis. Nat Chem Biol 16:302–309
39. Yan B, Ai Y, Sun Q et al (2021) Membrane damage during ferroptosis is caused by oxidation of phospholipids catalyzed by the oxidoreductases POR and CYB5R1. Mol Cell 81: 355–69 e10
40. Yan B, Ai Y, Zhang Z et al (2021) Assessing POR and CYB5R1 oxidoreductase-mediated oxidative rupture of PUFA in liposomes. STAR Protoc 2:100360
41. Zack GW, Rogers WE, Latt SA (1977) Automatic measurement of sister chromatid exchange frequency. J Histochem Cytochem 25:741–753
42. Wang H, Sun L, Su L et al (2014) Mixed lineage kinase domain-like protein MLKL causes necrotic membrane disruption upon phosphorylation by RIP3. Mol Cell 54:133–146

Chapter 8

LC-MS-Based Redox Phosphoipidomics Analysis in Ferroptosis

Wan-Yang Sun, Rong Wang, and Rong-Rong He

Abstract

Ferroptosis is a regulated form of cell death characterized by the accumulation of oxidized phospholipids, particularly oxidized phosphatidylethanolamines (PE), which serve as important biomarkers in the progression of various diseases. To facilitate the comprehensive investigation of ferroptosis in biological systems, we present a robust and versatile untargeted redox phospholipidomics method employing normal-phase liquid chromatography-mass spectrometry (LC-MS). This high-throughput technique enables the identification and quantification of dozens of oxidized phospholipid species in a single run, providing valuable insights into the molecular mechanisms underlying ferroptosis. It has been successfully applied to diverse biological samples, including human patients, animals, and cell cultures, and offers a powerful tool for investigating the roles of oxidized phospholipids in the development and progression of various diseases.

Key words Ferroptosis, Lipid peroxidation, Oxidized phospholipid, Liquid chromatography-mass spectrometry, Redox phospholipidomics

1 Introduction

Ferroptosis is an iron-dependent, lipid peroxidation-mediated mode of programmed cell death that has garnered significant attention due to its implications in various pathological conditions, including cancer, neurodegeneration, and ischemia-reperfusion injury [1, 2]. One of the hallmarks of ferroptosis is the accumulation of oxidized phospholipids, particularly oxidized phosphatidylethanolamines (PE), which serve as important biomarkers in the progression of the ferroptotic process [3]. Traditional lipid peroxidation detection methods, such as C11-BODIPY and Liperfluo, have been widely used to investigate lipid peroxidation in biological systems [4, 5]. However, these methods have certain limitations, including low specificity, lack of sensitivity, and inability to identify and quantify individual oxidized lipid species.

Guido Kroemer and Daolin Tang (eds.), *Ferroptosis: Methods and Protocols*, Methods in Molecular Biology, vol. 2712, https://doi.org/10.1007/978-1-0716-3433-2_8,

Given the limitations of traditional lipid peroxidation detection methods, there is a pressing need for advanced techniques to accurately detect oxidized phospholipids in biological samples. Liquid chromatography-mass spectrometry (LC-MS) has emerged as a powerful tool for the analysis of complex lipid mixtures, offering high sensitivity, specificity, and the ability to identify and quantify a wide range of lipid species. Given the limitations of traditional lipid peroxidation detection methods, there is a pressing need for advanced techniques to accurately detect oxidized phospholipids in biological samples. Liquid chromatography-mass spectrometry (LC-MS) has emerged as a powerful tool for the analysis of complex lipid mixtures, offering high sensitivity, specificity, and the ability to identify and quantify a wide range of lipid species. The untargeted redox phospholipidomics approach, which combines normal-phase LC-MS with bioinformatics tools for data analysis, offers a promising platform for the systematic investigation of oxidized phospholipids in the context of ferroptosis and other disease states [6]. By facilitating the identification and quantification of numerous oxidized phospholipid species in a single analysis, this method provides valuable insights into the molecular mechanisms underlying lipid peroxidation and its potential therapeutic implications [7–9].

2 Materials

2.1 Lipid Extraction

1. 0.75% KCl solution (*see* **Note 1**).
2. Chloroform.
3. Methanol.
4. Sterile glass test tubes.
5. Centrifuge.
6. Vortex mixer.
7. Nitrogen evaporator or vacuum centrifugal concentrator.

2.2 Phosphate Content Determination

1. Methanol.
2. 70% $HClO_4$ (Perchloric acid).
3. 1 mM NaH_2PO_4 (Sodium dihydrogen phosphate).
4. 2.5% $Na_2MoO_4 \cdot 2H_2O$ (Sodium molybdate dihydrate).
5. 10% Ascorbic acid.
6. Oil bath.
7. Boiling water bath.
8. Microplate reader.
9. 96-well plate.

2.3 Liquid Chromatography-Mass Spectrometry Analysis

1. Reference standard compounds: PA(18:1/18:1), PC(18:1/18:1), PE(18:1/18:1), PG(18:1/18:1), PI(18:1/18:1), PS(18:1/18:1), CL(16:0/18:2/18:2/20:4)
2. Internal standard compounds: PA(16:0-d31/18:1), PC(16:0-d31/18:1), PE(16:0-d31/18:1), PG(16:0-d31/18:1), PI(16:0-d31/18:1), PS(16:0-d31/18:1), CL(14:0/14:0/14:0/14:0).
3. Methanol.
4. Mobile phase: isopropanol/hexane/water (285:215:5, V/V/V, solvent A) and isopropanol/hexane/water (285:215:40, V/V/V, solvent B) with 10 mM ammonium formate.
5. LC-Q-Exactive MS system.
6. Luna Silica (2) column (3 μm, 150 × 2.0 mm) (Phenomenex).

3 Methods

3.1 Lipid Extraction Using Folch Method [10]

1. Collect five million cells or 10 mg of tissue sample or 300 μL of blood sample in a clean and sterile manner.
2. For cell samples:
   - Centrifuge the cells at 500× *g* for 5 min to remove the supernatant.
   - Add 1 mL of 0.75% KCl solution to the cell pellet and pre-cool the sample on ice.
   - Ultrasonicate the sample for 10 s, cool on ice for 20 s, and cycle five times. Ensure that all steps are carried out in a dark environment. Clean the ultrasonic probe with deionized water when changing the sample.
   - Centrifuge the sample at 12,000× *g* for 10 min to pellet cellular debris.
3. For tissue samples:
   - Add two steel balls to the tissue sample, then add 1 mL of 0.75% KCl.
   - Place the sample in a pre-cooled lead plate and grind it with a tissue grinder for 3 times, 30 s each time, with an interval of 10 s.
4. For blood samples:
   - Add 300 μL of blood sample to a clean tube and make up to 1 mL with 0.75% KCl.
5. Transfer the pretreated samples to a glass test tube.
6. Add 5 mL of chloroform:methanol (2:1, v/v) per 1 mL of sample.

7. Place the test tube on ice for 1 h and vortex every 10 min to fully extract the lipids in the sample, and then centrifuge at 1000× *g* for 10 min to separate the two phases. The above operations should be protected from light.
8. Transfer the upper layer liquid to another glass tube and repeat the extraction with 3 mL of chloroform:methanol (2:1, v/v).
9. Combine the two lower clear liquids, filter with a funnel, and collect the filtrate.
10. Evaporate the solvent using a nitrogen evaporator or a vacuum centrifugal concentrator to obtain the lipid extract. Store at − 80 °C before use.

3.2 Phosphate Content Determination

1. Add 200 μL of methanol for resuspension and transfer to an Eppendorf tube.
2. Centrifuge the lipid extract at 15,000× *g* for 10 min at 4 °C to remove any remaining debris, and transfer the supernatant to a new Eppendorf tube.
3. Take 20 μL of the sample and add 125 μL of 70% $HClO_4$ to the tube.
4. Heat the sample in an oil bath at 170 °C for 20–40 min until the sample becomes clear. If it is not clear, extend the heating time.
5. Prepare standard solutions for Pi concentration determination according to the following table (Table 1. *see* **Note 2**). Add 1 mM NaH_2PO_4 solvent at the end and mix well.
6. Cool the heated samples to room temperature and add 825 μL of deionized water and 125 μL of 2.5% $Na_2MoO_4•2H_2O$ to each sample.

Table 1
Standard solutions for Pi concentration

Pi Concentration	1 mM NaH_2PO_4	70% $HClO_4$	Deionized water	2.5% $Na_2MoO_4•2H_2O$	10% Ascorbic acid
0 μM	0 μL	125 μL	825 μL	125 μL	125 μL
4.16 μM	5 μL	125 μL	820 μL	125 μL	125 μL
12.5 μM	15 μL	125 μL	810 μL	125 μL	125 μL
20.8 μM	25 μL	125 μL	800 μL	125 μL	125 μL
29.2 μM	35 μL	125 μL	790 μL	125 μL	125 μL
37.5 μM	45 μL	125 μL	780 μL	125 μL	125 μL
54.2 μM	65 μL	125 μL	760 μL	125 μL	125 μL

7. Add 125 μL of 10% ascorbic acid to each sample and standard solution, and vortex.
8. Heat the samples in a boiling water bath for 10 min and cool on ice after heating.
9. Add 200 μL of each sample and standard solution to a 96-well plate, and measure the absorbance at 660 nm using a microplate reader.
10. Plot the absorbance values against the Pi concentrations to create a standard curve.
11. Determine the Pi concentration of the sample using the established standard curve. The Pi concentration in the sample can be calculated by extrapolation from the standard curve.

3.3 Liquid Chromatography-Mass Spectrometry Analysis

3.3.1 Preparation of Standard and Sample Solutions

1. Prepare a 20 μM reference standard stock solution containing PA(18:1/18:1), PC(18:1/18:1), PE(18:1/18:1), PG(18:1/18:1), PI(18:1/18:1), PS(18:1/18:1), and CL(16:0/18:2/18:2/20:4) by dissolving these compounds in methanol (*see* **Note 3**).
2. Prepare a 20 μM internal standard stock solution containing PA(16:0-d31/18:1), PC(16:0-d31/18:1), PE(16:0-d31/18:1), PG(16:0-d31/18:1), PI(16:0-d31/18:1), PS(16:0-d31/18:1), and CL(14:0/14:0/14:0/14:0) by dissolving these compounds in methanol (*see* **Note 3**).
3. Dilute the reference substance stock solution successively by factors of 3, 10, 30, 100, 300, and 1000, respectively, to obtain a series of concentration reference substance solutions.
4. For calibration curves, add 20 μL reference standard solution, 2 μL internal standard solution, and 18 μL methanol to a glass inner tube.
5. For samples, add an aliquot of sample solution containing 20 nmol phosphorus to a glass inner tube based on the phosphorus content of the sample, then add 2 μL internal standard solution, and finally use methanol to make up the volume to 40 μL.

3.3.2 Perform LC-MS Analysis)

Perform LC-MS analysis (Fig. 1) according to the following conditions.

Liquid Chromatography System

- Column: Luna Silica (2) column (3 μm, 150 × 2.0 mm) (Phenomenex).
- Flow rate: 0.2 mL/min.
- Column temperature: 35 °C.
- Mobile phase: Solvent A and B.
- Gradient elution schedule (Table 2).
- Injection volume: 5 μL.

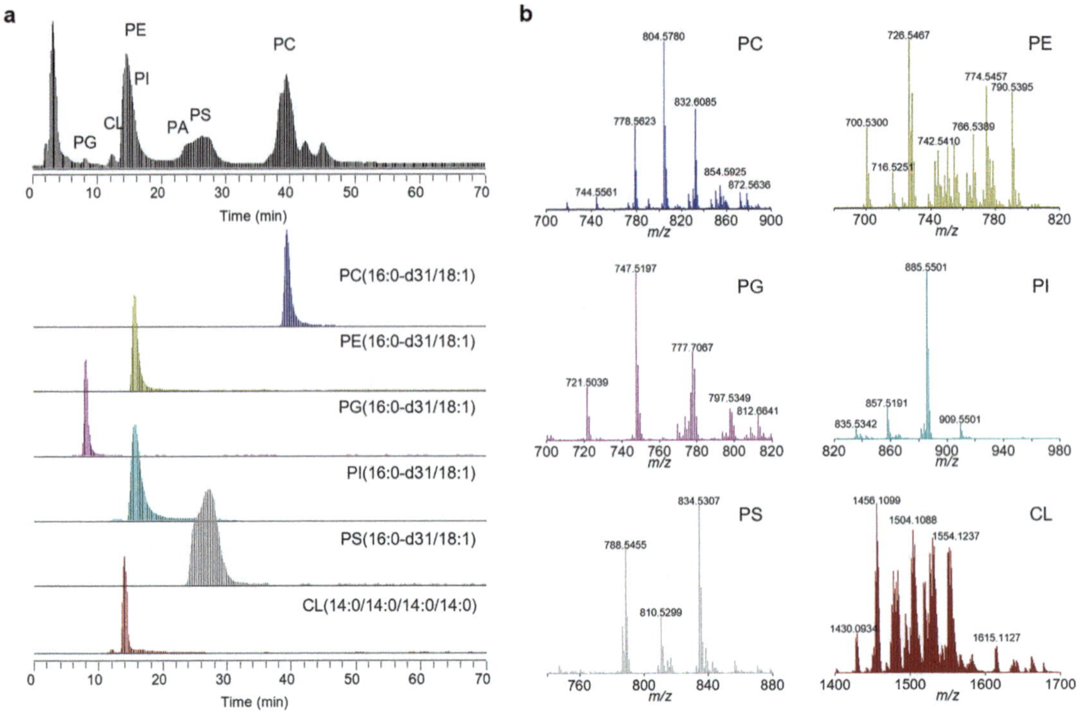

Fig. 1 Representative normal-phase LC-MS/MS chromatogram (**a**) and mass spectra for six major classes of phospholipids (**b**) in rat brain

Table 2
Gradient elution schedule

Time (min)	Mobile Phase A (%)	Mobile Phase B (%)
0	90	10
23	68	32
32	35	65
35	0	100
70	0	100

Mass Spectrometry System

Ionization source parameters:

- Ionization mode: Negative ionization.
- Capillary spray voltage: −2.8 kV.
- Capillary temperature: 350 °C.
- Aux gas heater temp: 320 °C.
- S-Lens Rf level: 60.

- Sheath gas: 30 arb.
- Aux gas: 15 arb.
- Sweep gas: 2 arb.

Data acquisition parameters:

- Acquisition mode: Full MS-dd-MS^2.
- Scan range: *m/z* 400–1800.

Full MS analysis:

- Resolution: 70,000.
- Maximum injection time: 200 ms using 1 microscan.

MS^2 Analysis:

- Inclusion list of oxidized phospholipids with interest.
- Resolution: 17,500.
- Maximum injection time: 500 ms using 1 microscan.
- Collision energy: 24 eV.
- Isolation window: 1.0 Da.

3.4 Data Analysis

3.4.1 Feature Extraction

To extract features from the LC-MS data, use MS-DIAL [11] with the following parameters:

- Retention time begin, 0 min.
- Retention time end, 70 min.
- Mass range begin, 400 Da.
- Mass range end, 1800 Da.
- Accurate mass tolerance (MS^1), 0.01 Da; MS^2 tolerance, 0.025 Da.
- Maximum charge number, 2.
- Smoothing method, linear weighted moving average.
- Smoothing level, 3.
- Minimum peak width, 10 scan.
- Minimum peak height, 10,000.
- Mass slice width, 0.1 Da.
- Sigma window value, 0.5.
- Keep isotope until, 0.5 Da.
- Accurate mass tolerance for identification, 0.01 Da for MS^1 and 0.05 for MS^2.
- Retention time tolerance for alignment, 2 min.
- MS^1 tolerance for alignment, 0.015 Da.

- Remove feature based on peak height fold-change, true.
- Sample max/blank average, 5.
- Replace true zero values with 1/10 of the minimum peak height over all samples, false.

3.4.2 Identification of Oxidized Phospholipid

1. Use MS-DIAL or a self-built oxidized phospholipid database to preliminarily annotate the oxygenation number of oxidized phospholipid products using exact mass, such as PE(38: 4) + 2O. The MS^1 tolerance is set at 5 ppm (*see* **Note 4**).
2. Confirm the presence of the non-oxidized precursor in the sample to ensure that the oxidized phospholipid product is a result of oxidation of the non-oxidized precursor and not a result of a different metabolic pathway or artifact.
3. Use the retention time of each type of phospholipid internal standard to filter the corresponding type of oxidation products. Usually the retention time tolerance is set at ±10% (*see* **Note 5**).
4. Perform MS^2 fragment ion assignment on the oxidized phospholipid product of interest using LIPID MAPS database [12] (https://www.lipidmaps.org/) to confirm the identification result (*see* **Note 6**) (Fig. 2).

*3.4.3 Semi-Quantitative Analysis of Oxidized Phospholipid (See **Note 7**)*

1. Establish a standard curve for different types of phospholipids using the peak area ratios of the reference substance and the internal standard in the reference solution of serial

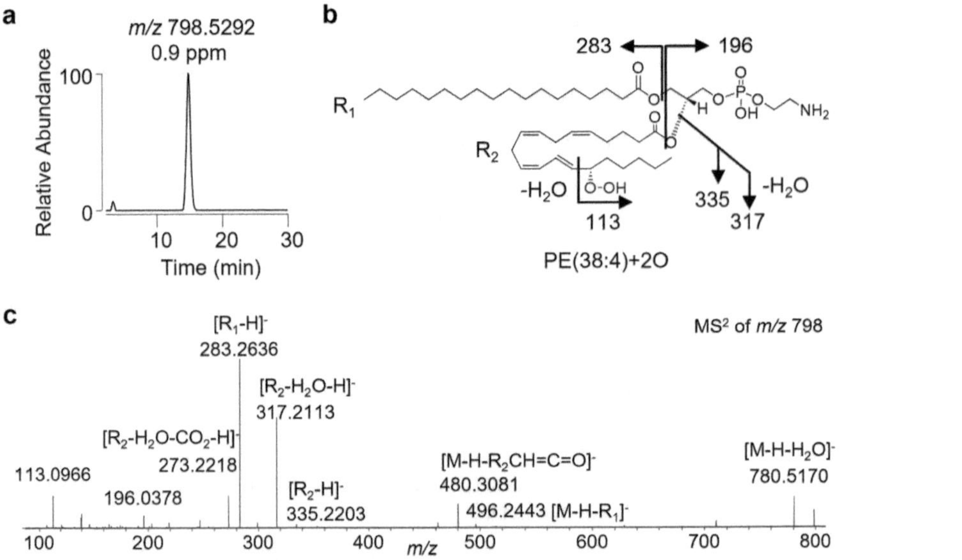

Fig. 2 Representative extract ion current (**a**), speculated fragmentations (**b**), and MS^2 spectrum (**c**) of PE(38: 4) + 20. The oxidized phospholipid is tentatively identified as 1-steraoyl-2-15-HpETE-*sn*-glycero-3-phosphatidylethanolamine

concentrations. This yield the concentration of phospholipids in the reference solution (C_{PL}) in micromoles (μM):

$$C_{PL} = a\tilde{n}A_{PL}/A_{IS} + b$$

where A_{PL} is the peak area of phospholipids in the reference solution, A_{IS} is the peak area of the corresponding internal standard, a is a constant factor, and b is a constant offset.

2. Use the standard curve to calculate the concentration of each oxidized phospholipid in the test solution. The peak area of each oxidized phospholipid in the test solution was divided by the peak area of the corresponding internal standard, multiplied by the constant factor (a), and added to a constant offset (b) that was determined from the standard curve. This yield the concentration of oxidized phospholipids (C_{oxPL}) in micromoles (μM):

$$C_{oxPL} = a\tilde{n}A_{oxPL}/A_{IS} + b$$

where A_{oxPL} is the peak area of each oxidized phospholipid in the test solution.

3. Normalize the concentration of each oxidized phospholipid in the sample by the amount of phosphorus or protein. The concentration of oxidized phospholipids is converted to picomoles per micromole of phosphorus (pmol/μmol Pi) using conversion factors appropriate for the specific sample:

$$C_{oxPL}\ (\text{pmol}/\mu\text{mol Pi}) = C_{oxPL}\ (\mu\text{M}) \cdot 40\ \mu\text{L}/20\ \text{nmol Pi}$$

4 Notes

1. The addition of 0.75% KCl to the solvent mixture is used to help separate the lipid phase from other components of the sample during lipid extraction.
2. When preparing the phosphate assay reagents, use freshly prepared solutions for optimal performance.
3. Avoid multiple freeze-thaw cycles for the lipid standards used in liquid chromatography-mass spectrometry analysis. Aliquot the standards to minimize freeze-thaw cycles, and store them at −20 °C or colder.
4. These annotations should be considered tentative until confirmed by other methods.
5. In normal-phase liquid chromatography, non-oxidized and oxidized phospholipids with the same head group have close elution times.

6. Verify the MS^2 fragmentation patterns by comparing them to those of standards or previously reported data, when available.
7. To calculate the content of oxidized phospholipids in a sample, the internal standard method should be employed and the results should be normalized by the amount of phosphorus.

References

1. Doll S, Proneth B, Tyurina YY et al (2017) ACSL4 dictates ferroptosis sensitivity by shaping cellular lipid composition. Nat Chem Biol 13:91–98
2. Dixon SJ, Lemberg KM, Lamprecht MR et al (2012) Ferroptosis: an iron-dependent form of nonapoptotic cell death. Cell 149:1060–1072
3. Kagan VE, Mao GW, Qu F et al (2017) Oxidized arachidonic and adrenic PEs navigate cells to ferroptosis. Nat Chem Biol 13:81–90
4. Kuhn H, Banthiya S, van Leyen K (2015) Mammalian lipoxygenases and their biological relevance. Biochim Biophys Acta 1851:308–330
5. Yamanaka K, Saito Y, Sakiyama J et al (2012) A novel fluorescent probe with high sensitivity and selective detection of lipid hydroperoxides in cells. RSC Adv 2:7894
6. Kagan VE, Tyurina YY, Sun WY et al (2020) Redox phospholipidomics of enzymatically generated oxygenated phospholipids as specific signals of programmed cell death. Free Radic Biol Med 147:231–241
7. Sun WY, Tyurin VA, Mikulska-Ruminska K et al (2021) Phospholipase iPLA(2)beta averts ferroptosis by eliminating a redox lipid death signal. Nat Chem Biol 17:465–476
8. Luo X, Gong HB, Gao HY et al (2021) Oxygenated phosphatidylethanolamine navigates phagocytosis of ferroptotic cells by interacting with TLR2. Cell Death Differ 28:1971–1989
9. Li W, Luo LX, Zhou QQ et al (2022) Phospholipid peroxidation inhibits autophagy via stimulating the delipidation of oxidized LC3-PE. Redox Biol 55:102421
10. Folch J, Lees M, Sloane Stanley GH (1957) A simple method for the isolation and purification of total lipides from animal tissues. J Biol Chem 226:497–509
11. Tsugawa H, Ikeda K, Takahashi M et al (2020) A lipidome atlas in MS-DIAL 4. Nat Biotechnol 38:1159–1163
12. Fahy E, Sud M, Cotter D et al (2007) LIPID MAPS online tools for lipid research. Nucleic Acids Res 35:W606–W612

Chapter 9

Monitoring Lysosome Function in Ferroptosis

Fangquan Chen, Rui Kang, Daolin Tang, and Jiao Liu

Abstract

Ferroptosis is a type of regulated cell death that occurs due to iron-induced membrane lipid peroxidation. Lysosomes, which are acidic, membrane-bound organelles containing various hydrolases, play a vital role in ferroptosis. They not only aid in the degradation of autophagic substrates, but also serve as signaling hubs in cell death. Specifically, lysosomes are involved in the induction and execution of ferroptosis through autophagy-mediated degradation of anti-ferroptotic proteins, lysosomal membrane permeability-mediated release of cathepsins, and iron-induced lysosomal membrane lipid peroxidation. Therefore, it is essential to have reliable methods for monitoring lysosomal functions, including lysosomal activity, pH, and membrane integrity, as well as iron accumulation and lipid peroxidation, to understand ferroptosis. This chapter introduces several protocols, such as western blotting, immunofluorescence, lysosomal probes, and lipid peroxidation assay kits, for monitoring the process of lysosome-related ferroptosis.

Key words Lysosome, Ferroptosis, Cell death, Autophagy

1 Introduction

Cell death can be classified into two categories: regulated cell death (RCD) and accidental cell death. RCD is characterized by controlled signaling mechanisms and a predictable rate of occurrence [1–3]. Ferroptosis is a type of non-apoptotic RCD that is triggered by iron-dependent lipid peroxidation on plasma membranes or membrane organelles [4]. Lysosomes play a critical role in sustaining cellular homeostasis by serving as centers for autophagic degradation and endocytic recycling [5]. Increased autophagy is generally a defense mechanism for cell survival in response to harmful stimuli, but excessive autophagy can damage cells or induce cell death by degrading protective proteins, both of which depend on lysosomes [6–9].

Selective autophagy-mediated degradation of anti-ferroptotic regulators promotes ferroptosis [10–16]. Specifically, ferritinophagy targets ferritin [17], lipophagy targets lipid droplets [18], clockophagy targets basic helix-loop-helix ARNT like 1 (BMAL1,

Guido Kroemer and Daolin Tang (eds.), *Ferroptosis: Methods and Protocols*, Methods in Molecular Biology, vol. 2712,
https://doi.org/10.1007/978-1-0716-3433-2_9,

also known as ARNTL) [19], and chaperone-dependent autophagy targets glutathione peroxidase 4 (GPX4) [14, 20]. On the other hand, lysosomal catabolism of albumin inhibits cystine deprivation-induced ferroptosis in a cathepsin B (CTSB)-dependent manner, which is a protective mechanism that can be enhanced by inhibiting mechanistic target of rapamycin complex 1 (MTOR1) [21]. However, damaged lysosomal membranes can also release CTSB to induce ferroptosis [22, 23].

Apart from their established role in degradation and recycling, lysosomes also play a central role in iron homeostasis and nutrient sensing [24]. Targeting lysosomes-related ferroptosis may provide new therapeutic options for ischemia-reperfusion injury, cancer, and neurodegenerative diseases [24, 25]. In this chapter, we summarize the methods to monitor lysosome activity and function in ferroptosis.

2 Materials

1. 1–16K high-speed freezing centrifuge (Sigma).
2. Phenylmethanesulfonyl fluoride (PMSF).
3. Vortexer and plate shaker.
4. Phosphate-buffered saline (PBS; Na_2HPO_4, KH_2PO_4, NaCl and KCl).
5. Bicinchoninic acid (BCA).
6. 6 × loading buffer (glycerol; tris; dithiothreitol; sodium dodecyl sulfate; bromophenol blue).
7. Skim milk.
8. Tris buffered saline buffer with tween 20 (TBST; 137 mM NaCl, 20 mM Tris, 0.1% Tween-20, pH 7.6).
9. Antibody diluents.
10. 6-well plates; 96-well plates; 24-well plate; 10 cm^2 petri dishes.
11. CO_2 cell incubator.
12. Image J software.
13. Image lab software.
14. Digital imaging systems.
15. Confocal cell culture dish.
16. Multifunctional imaging microplate reader.
17. 4% paraformaldehyde.
18. Pipette tips.
19. 0.25% trypsin-EDTA.
20. Fluorescence microscope (Zeiss).

21. Confocal fluorescence microscope (Nikon).
22. Trans-Blot Turbo transfer system.
23. C11-BODIPY.
24. LysoTracker Red DND-99.
25. LysoSenser pH indicators.
26. Acridine orange.
27. Fluorescent dextran.
28. DQ-BSA-Red.
29. 10 × electrophoretic buffer (30.3 g Tris-base; 144 g glycine; 10 g sodium dodecyl sulfate; add H_2O to 1 mL).
30. 10 × transmembrane buffer (30 g Tris-base; 144 g glycine; add H_2O to 1 mL).
31. Pipette.

3 Methods

3.1 Monitoring Lysosomal Integrity

Western blotting is a widely used technique for the detection of specific protein expression in cell or tissue extracts, particularly for proteins with low expression levels [26]. Generally, cells or tissues are lysed with a lysis buffer and boiled for denaturation, followed by transfer to nitrocellulose or polyvinylidene fluoride membranes via SDS-polyacrylamide gel electrophoresis (PAGE). The target protein is then recognized using specific primary antibodies, and the primary antibody is detected by direct conjugation with horseradish peroxidase (HRP) or alkaline phosphatase, using the principle of antigen-antibody reaction between different species. While several detection methods are available, chemiluminescent substrate-based detection is the most sensitive and straightforward approach, producing light as a by-product when combined with the enzyme. Finally, the chemiluminescence signal is captured by a digital imaging system. It is recommended to use western blotting to monitor nuclear or cytoplasmic CTSB protein levels during ferroptosis [22, 23].

3.1.1 Nuclear and Cytoplasmic Protein Extraction

The following is a detailed description of a modified extraction protocol, based on typical cell culture, from Beyotime Biotechnology.

1. Place the cells on ice, collect them by cell scraping, and centrifuge them to collect the cell precipitate. Discard the supernatant, making sure to avoid digesting the cells with trypsin to prevent degradation of the target protein to be extracted by trypsin.

2. Lyse the cells by using a mixture of cell plasma protein extraction reagent with phenylmethanesulfonyl fluoride (PMSF) at a concentration of 1 mM.
3. Vigorously vortex the cell precipitate for 5 s at maximum speed to ensure complete suspension and dispersion, and then place on ice for 10–15 min.
4. Add a 0.05 ratio of cytoplasmic protein extraction reagent, vortex vigorously for 5 s at the highest speed, and place the suspension on ice for 1 min.
5. Centrifuge the mixture at maximum speed of 5 s at 16,000 g for 5 min at 4 °C and collect the supernatant as cytoplasmic protein.
6. Discard the supernatant after centrifugation and add an appropriate amount of nuclear protein extraction reagent containing PMSF.
7. Vigorously vortex the cell precipitate at maximum speed for 30 s, then place on ice and repeat vortex every 2 min for a total of 30 min.
8. Centrifuge the mixture at 16,000 g for 10 min at 4 °C, and aspirate the supernatant as cellular nuclear protein for subsequent Western blotting.
9. Analyze the concentrations of cytoplasmic and nuclear protein supernatants using the bicinchoninic acid (BCA) assay from Thermo Fisher Scientific.

3.1.2 Western Blotting

1. Take 30 μg of protein sample and mix it completely with 6 × loading buffer (Bio-Rad).
2. Load proteins of the same mass onto SDS-PAGE gel wells, along with molecular weight markers.
3. Run the gel at 80 V for 30 min, then increase the voltage to 120 V and complete the run in about 60 min.
4. Transfer the protein from SDS-PAGE gel to polyvinylidene fluoride membrane at a constant current of 240 mA by wet transfer method.
5. Wash the membranes with phosphate buffered saline (PBS; Na_2HPO_4, KH_2PO_4, NaCl, and KCl) for 5 min and transfer to 50 mL of 5% skim milk in Tris buffered saline buffer with tween 20 (TBST; 137 mM NaCl, 20 mM Tris, 0.1% Tween-20, pH 7.6) for 1 h at room temperature.
6. Wash 1 time for 5 min with an appropriate amount of TBST.
7. Incubate the membrane with an appropriate amount of primary antibody (diluted 1:1000–1:2000) and shake gently overnight at 4 °C.

8. Wash 5 times for 5 min each with an appropriate amount of TBST.
9. Incubate the membrane with an appropriate amount of species-appropriate HRP-linked secondary antibody (diluted 1:5000) and shake gently for 1 h at room temperature.
10. Wash 5 times for 5 min each with an appropriate amount of TBST.
11. Configure the chemiluminescent substrate according to the instructions and subsequently apply uniformly to the membrane.
12. Capture chemiluminescence signals using digital imaging systems (Bio-Rad).
13. Quantify target protein expression levels using image analysis software.

3.2 Monitoring Lysosomal Membrane Permeabilization

Lipid peroxidation-induced membrane rupture is a crucial event in ferroptosis that affects not only the plasma membrane but also various organelle membranes, including the lysosomal membrane. Lysosomal membrane permeabilization (LMP) increases and releases hydrolases from the lysosomal lumen into the cytosol, which can result in cell death either alone or in combination with other forms of cell death. Therefore, there are different methods available for the functional quantification and visualization of LMP during ferroptosis, such as (1) direct visualization of LMP by monitoring the release of fluorescent dextran from lysosomes into the cytosol [27], (2) detection of galectin translocation from cytoplasm to damaged lysosomes [28], and (3) assay of acridine orange uptake and intracellular redistribution [29].

3.2.1 Fluorescent Dextran

In healthy cells, fluorescent dextrose is typically found in punctate structures within the lysosomes. However, after LMP occurs, diffuse staining patterns are observed throughout the cytoplasm. To determine the size of the pores formed by the membrane during LMP, the release of dextran of different sizes (10, 70, and 500 kDa) in different colors is compared.

1. Cells are seeded in a culture dish and allowed to adhere.
2. Fluorescent dextran is added to the medium in the concentration of 100 μg/ml, such as Alexa Fluor® 488- or 594-Dextran (10 kDa) or fluorescein isothiocyanate (FITC)-Dextran (70 kDa).
3. The cells are incubated for 3–16 h, but the optimal dextran concentration and incubation time vary by cell line and should be optimized.
4. After dextran loading, the cells are washed 3 times in PBS and re-incubated with fresh medium for 2 h, after which the entry

of dextran into the lysosomes can be viewed using an inverted fluorescence microscope.

5. Cells loaded with dextran are subsequently treated with ferroptosis inducers (e.g., RSL3 and erastin).
6. After incubation for the appropriate time, the degree of cellular LMP is observed using fluorescence confocal and quantified using Image J software. It should be noted that if the degree of drug-induced LMP is weak, it may be difficult to visualize by fluorescence microscopy.

3.2.2 Galectin Puncta Assay

Galactins are soluble carbohydrate-binding lectins that can bind β-galactoside and a wide range of complex carbohydrates. Among them, galectin-1 and -3 translocate from the cytosol to the lysosomal membrane upon the permeabilization of the lysosomal membrane [30, 31]. This allows the development of a visualization of staining pattern changes by immunofluorescence, i.e., changes from diffuse cytosolic staining to punctate staining. However, co-staining with lysosomal membrane proteins such as LAMPs is required to ensure that galactose lectins are indeed localized to the lysosome. The following is a detailed experimental procedure:

1. Seed cells in a culture dish and allow cells to adhere so that they reach 65% confluence at the time of treatment.
2. Administer drug treatment when the cells are fully adhered (usually the next day). Set up a positive control such as L-leucyl L-leucine methyl ester treatment for 4 h (2 mM).
3. Remove the medium and fixate cells in 4% paraformaldehyde for 15 min at room temperature.
4. Wash the cells three times with PBS for 5 min each.
5. Incubate the cells in the blocking buffer solution (Cell Signaling Technology; #12411) at room temperature for 1 h.
6. Discard the blocking solution and wash the cells three times with PBS for 5 min each.
7. Dilute galectin-1 (Cell Signaling Technology; 12,936; 1:50) or galectin-3 antibodies (Cell Signaling Technology; 87,985; 1: 400) in fluorescent antibody diluent (Cell Signaling Technology; 12,378) and incubate the cells with antibody solutions overnight at 4 °C in a humidified chamber.
8. Wash the cells three times with PBS for 5 min each.
9. Dilute the fluorescein-Alexa Fluor®488/594-conjugated secondary antibodies of different species with dilution buffer and incubate the cells for 2 h away from light.
10. Rinse the cells with 1 × PBS three times for 5 min each, away from light.

11. Incubate the cells for 30 min with Hochst 33,342 (1:2000) to label nuclei.
12. Wash the cells twice with PBS.
13. Add an appropriate amount of anti-quenching agent.
14. Observe the cells under a fluorescent confocal microscope.
15. Analyze the number of galectin dots per cell and the percentage of cells positive for galectin dots using Image J software.

3.2.3 Acridine Orange Staining

Acridine orange (AO) is a lysosomal metachromatic fluorescent dye that can be used to assess lysosomal membrane permeability. In intact lysosomes, AO is a protonated oligomer and emits red fluorescence (excitation wavelength 555 nm, emission wavelength 617 nm). In the cytoplasm, it is a monomer deprotonated form and emits green fluorescence (excitation wavelength 490 nm, emission wavelength 528 nm). After entering lysosomes, AO undergoes redistribution, which allows for the analysis of lysosomal membrane permeability through changes in fluorescence patterns.

1. Prepare cells to be tested in a 96-well plate or slide culture dish.
2. Carefully remove the cell culture medium.
3. Carefully add 500 μL of pre-warmed cleaning solution at 37 °C along the well wall of the cell culture well, ensuring the surface of the culture well is covered.
4. Carefully remove the cleaning fluid.
5. Repeat **step 3**.
6. Carefully add 5 μL of staining solution to the cell culture well along the well wall, covering the surface of the culture well.
7. Incubate the cell in a 37 °C cell incubator for 15 min while avoiding light.
8. Remove the staining solution.
9. Carefully add 500 μL of pre-warmed cleaning solution at 37 °C along the well wall of the cell culture well, ensuring the surface of the culture well is covered.
10. Carefully remove the cleaning fluid.
11. Repeat **steps 9** and **10** once.
12. Carefully add 500 μL of pre-warmed cleaning solution at 37 °C along the well wall of the cell culture well.
13. Observe using a fluorescence confocal microscope (excitation wavelength 555 nm, emission wavelength 617 nm – red fluorescence decreases, indicating enhanced LMP) or detect using a fluorescence microplate reader (excitation wavelength 490 nm, emission wavelength 528 nm – elevated RFU, indicating enhanced LMP).

3.3 Monitoring Lysosomal Activity

Autophagy promotes ferroptosis, particularly through the degradation of ferritin, known as ferritinophagy, which occurs mainly via lysosomal-dependent degradation. Degradation of ferritin increases the concentration of labile iron within the cell, leading to lipid peroxidation-dependent ferroptosis. The inhibition of lysosomal activity using inhibitors, such as bafilomycin A1 or hydroxychloroquine, has been shown to significantly decrease ferroptosis. To quantify lysosomal activity, a self-quenching reporter substrate, DQ-BSA-Red, is used. DQ-BSA-Red is cleaved by lysosomal proteases upon entering the lysosome, resulting in the release of the fluorescent fragment and the subsequent emission of bright fluorescence. Therefore, DQ-BSA-Red can be used as an indicator of proper lysosomal function [32].

1. Cells are seeded in a culture dish and allowed to adhere so that they reach 65% confluency at the time of treatment.
2. The next day, the cells are washed three times with incomplete medium (without fetal bovine serum).
3. DQ-BSA-Red is diluted in complete medium (without 10% fetal bovine serum) to a final concentration of 10 μg/mL. Then, 200 μL is added to cover the petri dish.
4. Cells are placed in a 5% CO_2 incubator at 37 °C for 1 h.
5. After incubation, cells are washed three times with incomplete medium.
6. The cells are then incubated in incomplete medium for 2 h in a 5% CO_2 incubator at 37 °C.
7. After incubation, the cells are treated with ferroptosis drugs for the appropriate time.
8. Following treatment, the drug is discarded, and cells are washed with PBS three times.
9. Cells are fixed with 4% paraformaldehyde for 15 min at room temperature.
10. Cells are washed three times with PBS.
11. The nuclei are stained with 1 mg/ml DAPI in sterile water and incubated at room temperature for 30 min.
12. Cells are washed three times with PBS.
13. DQ-BSA fluorescence is analyzed and quantified using Image J software by visualizing cells with confocal microscopy.

3.4 Monitoring Lysosomal pH

LMP not only releases CTSB to mediate ferroptosis but also causes a large amount of H^+ leakage, accelerating the death process. Therefore, lysosomal pH serves as an indicator of lysosome-dependent ferroptosis. LysoSenser pH indicators, commercially available fluorescent dyes, selectively stain acidic compartments in living cells [33, 34]. This class of probes can selectively label acidic organelles and aggregate on spherical organelles. In a neutral environment, the probe produces blue fluorescence (Ex/Em = 329 nm/440 nm), but emits yellow fluorescence (Ex/Em = 384 nm/540 nm) when the environment becomes acidic.

1. Seed cells in a special culture dish and allow the cells to adhere.
2. Treat the cells with ferroptosis drugs for the appropriate time.
3. Remove the reagent from −20 °C to rewarm and centrifuge to concentrate the DMSO solution at the bottom of the tube.
4. Dilute the 1 mM probe stock solution to the final working concentration in the growth medium or PBS. It is recommended to use at least 1 μM of working concentration for the LysoSensor probes (*see* **Note 1**).
5. Discard the culture medium and add the appropriate amount of 37 °C pre-warmed working solution containing the probe. Incubate under growth conditions for 30 min to 2 h, depending on the cell type.
6. Discard the culture medium and wash the cells three times with PBS.
7. Replace the loading solution with fresh medium and observe the cells using a fluorescent confocal microscopy fitted with the correct filter set.

3.5 Monitoring Lysosomal Reactive Oxygen Species

Acidic organelles, such as endosomes and lysosomes, play a crucial role in iron uptake and release. The free iron released by these organelles is involved in local reactive oxygen species (ROS) production through the Fenton reaction. On the other hand, lysosomal inhibitors such as bafilomycin A1 can significantly impede ferroptosis and ROS production. For evaluating lipid peroxidation and antioxidant effects in model membrane systems and living cells, C11-BODIPY is a commonly used, lipid-soluble, ratiometric fluorescent indicator. Under the oxidation of living cells, the excitation maximum shifts from 581 nm to 500 nm, and the emission maximum shifts from 591 nm to 510 nm [35, 36]. Therefore, C11-BODIPY is a widely used product for ferroptosis research. Similarly, LysoTracker® Red DND-99 is a red fluorescently labeled lysosomal probe with a maximum excitation/emission wavelength of 577/590 nm, commonly used to label lysosomes [37, 38]. These two fluorescent probes can label the ROS fluorescence signal of lysosomes during ferroptosis.

The following protocol outlines the use of a modified Image-iT lipid peroxidation kit and LysoTracker™ Red DND-99 from Thermo Fisher Scientific:

1. Seed the appropriate number of cells in culture dishes and incubate overnight at 37 °C in 5% CO_2.

2. Treat the cells with the drug to be studied and incubate for the desired time.
3. Allow the cryopreserved C11-BODIPY and LysoTracker™ Red DND-99 to thaw at room temperature for 10 min.
4. Add 10 μM of C11-BODIPY and 1 μM of LysoTracker™ Red DND-99 to the cells and incubate at 37 °C for 30 min (*see* **Note 2**).
5. Remove the medium and rinse the cells with PBS three times.
6. Use fluorescence microscopy to determine fluorescence. Reduced dyes can be visualized with an excitation/emission of 581/591 nm (Texas Red filter), while oxidized dye can be visualized with an excitation/emission of 488/510 nm (FITC filter).
7. Stain the nuclei with Hoechst 33342 (5 μg/mL) in the dark at 37 °C for 20 min.
8. Remove the Hoechst 33342 solution and rinse the cells with PBS twice.
9. Use the NucBlue channel to identify cells and observe fluorescence co-localization of lysosomes with C11-BODIPY.
10. Quantify fluorescence intensity using ImageJ software.

4 Notes

1. The optimal working concentration and incubation time should be optimized for different experimental requirements, cell types, membrane permeability of cells or tissues, etc. To reduce possible false positives caused by excessive probe loading, it is recommended to use as low a concentration as possible without compromising the staining effect. In addition, if cells are incubated in dye-free medium after staining, attenuation of the fluorescent signal and vacuolation of the cells will be observed.
2. The appropriate working concentration of the storage solution should be determined based on the specific cell type and experimental objectives. To achieve the desired working concentration, the storage solution can be diluted with either medium or PBS buffer.

Acknowledgments

Research by J.L. was supported by grants from the National Natural Sciences Foundation of China (82070613).

References

1. Tang D, Kang R, Berghe TV et al (2019) The molecular machinery of regulated cell death. Cell Res 29:347–364
2. Xie Y, Hou W, Song X et al (2016) Ferroptosis: process and function. Cell Death Differ 23: 369–379
3. Liu J, Kang R, Tang D (2022) Signaling pathways and defense mechanisms of ferroptosis. FEBS J 289:7038–7050
4. Liu J, Kang R, Tang D (2021) Signaling pathways and defense mechanisms of ferroptosis. FEBS J 289:7038–7050
5. Kroemer G, Jaattela M (2005) Lysosomes and autophagy in cell death control. Nat Rev Cancer 5:886–897
6. Chen F, Cai X, Kang R, et al (2023) Autophagy-Dependent Ferroptosis in Cancer. Antioxid Redox Signal 39(1–3):79–101
7. Bialik S, Dasari SK, Kimchi A (2018) Autophagy-dependent cell death – where, how and why a cell eats itself to death. J Cell Sci 131:jcs215152
8. Denton D, Kumar S (2019) Autophagy-dependent cell death. Cell Death Differ 26: 605–616
9. Liu J, Kuang F, Kroemer G et al (2020) Autophagy-dependent ferroptosis: machinery and regulation. Cell Chem Biol 27:420–435
10. Chen X, Song X, Li J et al (2022) Identification of HPCAL1 as a specific autophagy receptor involved in ferroptosis. Autophagy 19:54–74
11. Li J, Liu J, Xu Y et al (2021) Tumor heterogeneity in autophagy-dependent ferroptosis. Autophagy 17:3361–3374
12. Yang L, Ye F, Liu J et al (2022) Extracellular SQSTM1 exacerbates acute pancreatitis by activating autophagy-dependent ferroptosis. Autophagy 19:1733–1744
13. Song X, Zhu S, Chen P et al (2018) AMPK-mediated BECN1 phosphorylation promotes ferroptosis by directly blocking system Xc(−) activity. Curr Biol 28:2388–99 e5
14. Liu Y, Wang Y, Liu J et al (2021) Interplay between MTOR and GPX4 signaling modulates autophagy-dependent ferroptotic cancer cell death. Cancer Gene Ther 28:55–63
15. Liu J, Liu Y, Wang Y et al (2022) TMEM164 is a new determinant of autophagy-dependent ferroptosis. Autophagy 19:945–956
16. Chen F, Cai X, Kang R et al (2023) Autophagy-dependent ferroptosis in cancer. Antioxid Redox Signal
17. Hou W, Xie Y, Song X et al (2016) Autophagy promotes ferroptosis by degradation of ferritin. Autophagy 12:1425–1428
18. Bai Y, Meng L, Han L et al (2019) Lipid storage and lipophagy regulates ferroptosis. Biochem Biophys Res Commun 508:997–1003
19. Yang M, Chen P, Liu J et al (2019) Clockophagy is a novel selective autophagy process favoring ferroptosis. Sci Adv 5:eaaw2238
20. Wu Z, Geng Y, Lu X et al (2019) Chaperone-mediated autophagy is involved in the execution of ferroptosis. Proc Natl Acad Sci U S A 116:2996–3005
21. Armenta DA, Laqtom NN, Alchemy G et al (2022) Ferroptosis inhibition by lysosome-dependent catabolism of extracellular protein. Cell Chem Biol 29:1588–600.e7
22. Kuang F, Liu J, Li C et al (2020) Cathepsin B is a mediator of organelle-specific initiation of ferroptosis. Biochem Biophys Res Commun 533:1464–1469
23. Nagakannan P, Islam MI, Conrad M et al (2021) Cathepsin B is an executioner of ferroptosis. Biochim Biophys Acta Mol Cell Res 1868:118928
24. Dielschneider RF, Henson ES, Gibson SB (2017) Lysosomes as oxidative targets for cancer therapy. Oxid Med Cell Longev 2017: 3749157
25. Mai TT, Hamai A, Hienzsch A et al (2017) Salinomycin kills cancer stem cells by sequestering iron in lysosomes. Nat Chem 9:1025–1033
26. Towbin H, Staehelin T, Gordon J (1979) Electrophoretic transfer of proteins from polyacrylamide gels to nitrocellulose sheets: procedure and some applications. Proc Natl Acad Sci U S A 76:4350–4354
27. Aits S, Jaattela M, Nylandsted J (2015) Methods for the quantification of lysosomal membrane permeabilization: a hallmark of lysosomal cell death. Methods Cell Biol 126: 261–285
28. Aits S, Kricker J, Liu B et al (2015) Sensitive detection of lysosomal membrane permeabilization by lysosomal galectin puncta assay. Autophagy 11:1408–1424
29. Chen F, Zhu S, Kang R et al (2022) ATP6V0D1 promotes alkaliptosis by blocking STAT3-mediated lysosomal pH homeostasis. Cell Rep 42:111911
30. Paz I, Sachse M, Dupont N et al (2010) Galectin-3, a marker for vacuole lysis by invasive pathogens. Cell Microbiol 12:530–544

31. Barondes SH, Castronovo V, Cooper DN et al (1994) Galectins: a family of animal beta-galactoside-binding lectins. Cell 76:597–598
32. Hu M, Li P, Wang C et al (2022) Parkinson's disease-risk protein TMEM175 is a proton-activated proton channel in lysosomes. Cell 185:2292–308.e20
33. Kang JS, Kostov Y (2002) Ratiometric pH measurements using LysoSensor DND-192. J Biochem Mol Biol 35:384–388
34. Matthaiou EI, Chiu W, Conrad C et al (2022) Macrophage Lysosomal Alkalinization Drives Invasive Aspergillosis in a Mouse Cystic Fibrosis Model of Airway Transplantation. J Fungi (Basel, Switzerland) 8:751
35. Drummen GP, van Liebergen LC, Op den Kamp JA et al (2002) C11-BODIPY (581/591), an oxidation-sensitive fluorescent lipid peroxidation probe: (micro)spectroscopic characterization and validation of methodology. Free Rad Biol Med 33:473–490
36. Pap EH, Drummen GP, Post JA et al (2000) Fluorescent fatty acid to monitor reactive oxygen in single cells. Methods Enzymol 319: 603–612
37. Lemieux B, Percival MD, Falgueyret JP (2004) Quantitation of the lysosomotropic character of cationic amphiphilic drugs using the fluorescent basic amine Red DND-99. Anal Biochem 327:247–251
38. Neun BW, Stern ST (2011) Monitoring lysosomal activity in nanoparticle-treated cells. Methods Mol Biol (Clifton, NJ) 697:207–212

Chapter 10

Monitoring Mitochondria Function in Ferroptosis

Fangquan Chen, Jiao Liu, Daolin Tang, and Rui Kang

Abstract

Ferroptosis is a type of regulated necrosis driven by uncontrolled membrane lipid peroxidation. Mitochondria, which are membrane-bound organelles present in almost all eukaryotic cells and play a central role in energy metabolism and various types of cell death, have a complicated role in ferroptosis. On one hand, mitochondrial-derived iron metabolism and reactive oxygen species (ROS) production may promote ferroptosis. On the other hand, mitochondria also possess a dihydroorotate dehydrogenase (DHODH)-dependent antioxidant system that detoxifies lipid peroxides. This chapter summarizes several methods, such as western blotting, immunofluorescence, cell viability assays, mitochondrial fluorescent probes, adenosine 5′-triphosphate (ATP) assay kits, mitochondrial respiration, and mitophagy tests, that may enable researchers to gain a deeper understanding of the dual role of mitochondria in ferroptosis.

Key words Mitochondria, Ferroptosis, Methods

1 Introduction

Cellular life activities require the supply of various nutrients, including the trace element iron [1]. However, iron overload induces lipid peroxidation reactions and cell death, especially ferroptosis [2]. Unlike other known regulated cell deaths, such as apoptosis, necroptosis, and pyroptosis, ferroptosis doesn't require traditional cell death drivers, such as caspases, gasdermin D (GSDMD), and mixed-lineage kinase domain-like protein (MLKL) [3]. Instead, toxic lipid metabolism production, such as 4-hydroxynonenal (4-HNE), can drive ferroptosis [4]. Impaired ferroptosis is involved in multiple pathological conditions and has become an attractive target to treat diseases [5].

Although the mechanisms of ferroptosis remain largely unknown, the process is involved in multiple metabolism abnormalities [4]. Since mitochondria, a double-membrane organelle, occupy a central position in cellular metabolism, dysfunctional mitochondria are involved in the control of ferroptosis sensitivity [6, 7]. Morphologically, ferroptotic cells have smaller

Guido Kroemer and Daolin Tang (eds.), *Ferroptosis: Methods and Protocols*, Methods in Molecular Biology, vol. 2712, https://doi.org/10.1007/978-1-0716-3433-2_10,

mitochondria, increased membrane density, reduced or absent mitochondrial crista, and mitochondrial outer membrane breaks [8]. In many cases, increased iron accumulation, ROS production, and the tricarboxylic acid (TCA) cycle in mitochondria can promote ferroptosis [9–13]. However, some evidence contradicts the view that mitochondria support ferroptosis. For example, an early study showed that cells lacking mitochondrial DNA ($\rho 0$ cells) exhibited the same sensitivity to ferroptosis as parental cells [14]. Additionally, the TCA cycle is associated with the production of many antioxidants. In particular, the dihydroorotate dehydrogenase (DHODH)–CoQH2/ubiquinone antioxidant system plays a major role in mitochondria to inhibit ferroptosis [15]. Thus, the role of mitochondria in ferroptosis is highly context-dependent.

This chapter presents various techniques for monitoring mitochondrial function in adherent mammalian cell cultures during ferroptosis.

2 Materials

1. High-speed freezing centrifuge.
2. Phenylmethanesulfonyl fluoride (PMSF).
3. Vortexer and plate shaker.
4. Phosphate-buffered saline (PBS; Na_2HPO_4, KH_2PO_4, NaCl and KCl).
5. Bicinchoninic acid (BCA).
6. 6 × loading buffer (glycerol; tris; dithiothreitol; sodium dodecyl sulfate; bromophenol blue).
7. Skim milk.
8. Tris buffered saline buffer with tween 20 (TBST; 137 mM NaCl, 20 mM Tris, 0.1% Tween-20, pH 7.6).
9. 6-well plates; 96-well plates; 10 cm^2 petri dishes.
10. CO_2 cell incubator.
11. Image J software.
12. Image Lab software.
13. Digital imaging systems.
14. Confocal cell culture dish.
15. Multifunctional imaging microplate reader.
16. Pipette tips.
17. Fluorescence microscope (Zeiss).
18. Confocal fluorescence microscope (Nikon).
19. 10 × electrophoretic buffer (30.3 g Tris-base; 144 g glycine; 10 g sodium dodecyl sulfate; add H_2O to 1 mL).

20. 10 × transmembrane buffer (30 g Tris-base; 144 g glycine; add H_2O to 1 mL).
21. Pipette.
22. RIPA lysis buffer [50 mM Tris (pH 7.4); 150 mM NaCl; 1% TritonX-100; 1% sodium deoxycholate; 0.1% sodium dodecyl sulfate; 2 mM sodium pyrophosphate; 25 mM β-glycerophosphate; 1 mM EDTA; 1 mM Na_3VO_4; 0.5 μg/ml leupeptin].
23. Ultrasonicator.
24. 1.5 and 2 mL centrifuge tube.
25. Oligomycin; FCCP [carbonylcyanide phenylhydrazone, carbonyl cyanide 4-(trifluoromethoxy) phenylhydrazone]; rotenone; antimycin A.
26. Opaque 96-well plates.
27. 2-DG.
28. Tetramethylrhodamine, methyl ester (TMRM) and JC-1.
29. 1× Live Cell Imaging Buffer (LCIS).
30. Cell counting kit-8 (CCK-8).
31. 4 °C refrigerator.
32. MitoSOX.
33. MitoTracker red.
34. mRFP-GFP-LC3B fusion protein adenovirus.
35. Trans-Blot turbo transfer system.

3 Methods

3.1 Monitoring Mitochondrial Proteins

Western blotting is an essential technique for evaluating protein levels during ferroptosis, and several marker proteins have been identified for this purpose. These include acyl-CoA synthetase long-chain family member 4 (ACSL4) [16], DHODH [15], GPX4 [17], transferrin receptor (TFRC) [18], and CDGSH iron sulfur domain 1 (CISD1) [12], and decorin (DCN) [19].

Protein extraction and western blotting are optimized for performing according to Cell Signaling Technology (*see* **Note 1**).

3.1.1 Protein Extraction

1. Discard the cell culture medium.
2. Wash the cells twice with 2 mL of pre-cooled PBS (pH 7.4; NaCl 8.0 g; KCl 0.2 g; Na_2HPO_4 1.44 g; KH_2PO_4 0.24 g; add ddH_2O to 1000 mL).
3. Add 10 μL of phenylmethanesulfonyl fluoride (PMSF; 100 mM) to 1 mL of lysis buffer and shake well on ice. Ensure

that the PMSF is shaken well until no crystals are present before mixing with the lysis buffer.

4. Add 400 μL of PMSF-containing lysate to each cell bottle and scrape the cells to collect them.
5. Place the cells on ice and vortex vigorously for 5 s every 10 min, then lyse for 30 min.
6. Use an ultrasonicator to break the nucleic acids and centrifuge the lysate at 15,000 g for 5 min at 4 °C.
7. Transfer the centrifuged supernatant to a 2 mL centrifuge tube.
8. Determine the protein concentration using the bicinchoninic acid (BCA) assay (Thermo Fisher Scientific).

3.1.2 Western Blotting

1. Take 30 μg of protein solution as the sample volume and add 4× SDS sample buffer to a final concentration of 1×. The total volume of the sample should generally not exceed 15 μL, but the maximum limit of the sample wells can accommodate up to 20 μL of sample.
2. Load the protein molecular weight marker and samples (ensuring the same total mass per well) into the SDS-PAGE well.
3. Run the gel at 80 V for 30 min, then increase the voltage to 120 V. Stop the electrophoresis when bromophenol blue has run out.
4. After electrophoresis, transfer the SDS-PAGE gel to a PVDF membrane at 240 mA for 70 min.
5. After transfer, wash the membrane once with PBS.
6. Block the membrane with 5% skim milk in TBST (Tris-buffered saline buffer with Tween 20) for 1 h at room temperature.
7. Wash the membrane twice with TBST.
8. Incubate the membrane overnight at 4 °C with a primary antibody (diluted according to the product's recommended ratio) and shake gently.
9. Wash the membrane 3 times for 10 min each with TBST.
10. Incubate the membrane with a species-appropriate HRP-linked secondary antibody (diluted 1:5000) and shake gently for 1 h at room temperature.
11. Wash the membrane 3 times for 10 min each with TBST.
12. Prepare the chemiluminescent substrate according to the instructions and apply it uniformly to the membrane.
13. Capture the signal using a chemiluminescence digital imaging system.
14. Quantify the target protein expression levels using image analysis software.

3.2 Monitoring Energy Production

Normal cells require a balanced supply of energy and nutrients to maintain their survival, and any disruption to this balance can result in metabolic stress. When faced with metabolic stress, AMP-activated protein kinase (AMPK), a critical intracellular energy state receptor, is activated to restore energy homeostasis [20]. However, if metabolic stress persists or is overactivated, it can induce cell death. Mitochondria, the central organelles of oxidative phosphorylation and energy metabolism, plays an essential role in sensing and regulating energy homeostasis. Furthermore, adenosine 5′-triphosphate (ATP) can act as a damage-associated molecular pattern (DAMP) molecule and mediate various biological processes in cell death [21]. Therefore, we recommend using ATP kits and Seahorse Cellular Energy Metabolism Analysis System to assess mitochondrial function in ferroptotic cells [22].

3.2.1 ATP Assay

ATP plays a crucial role in multiple physiological and pathological processes within cells. A decrease in ATP levels usually indicates impaired or reduced mitochondrial function. To detect ATP levels, a kit was developed based on the principle that firefly luciferase requires ATP to catalyze the production of fluorophore light. In the presence of excess firefly luciferase and fluorophore, the amount of fluorophore produced is proportional to the concentration of ATP within a specific concentration range. This kit offers a reliable and sensitive method to quantify ATP levels within cells.

1. Discard the cell culture medium and wash the cells with phosphate-buffered saline (PBS).
2. Add 200 μL of lysis buffer per well of a 6-well plate, and use a pipette to blow or shake the plate repeatedly to ensure complete cell lysis. Collect the cell lysates in 1.5 mL centrifuge tubes and keep them on ice throughout the process.
3. After lysis, centrifuge the samples at 4 °C for 5 min at 12,000 g and carefully remove the supernatant for further analysis.
4. Dilute the ATP standard solution in ATP detection lysis buffer to obtain an appropriate concentration range, depending on the expected ATP levels in the samples.
5. Prepare the ATP assay working solution by mixing 100 μL of ATP assay reagent with 400 μL of ATP assay reagent diluent. Ensure that all reagents are melted on ice before use.
6. Add 100 μL of the ATP assay working solution to each assay well or tube, and incubate for 3–5 min at room temperature to allow consumption of the background ATP and reduce background noise.
7. Add 20 μL of sample or standard to the assay well or tube, mix quickly with a pipette gun, and measure the relative light units (RLUs) or counts per minute (CPM) using a

chemiluminescence meter or a liquid scintillation counter after an interval of at least 2 s.

8. Calculate the ATP concentration in the sample from the standard curve generated by plotting the known concentrations of ATP standards against their corresponding RLU or CPM values.
9. Determine the protein concentration of the sample using the BCA assay, and convert the ATP concentration to nmol/mg protein.
10. The resulting ATP and protein concentrations can be used to normalize the ATP levels to the amount of protein in the sample.

3.2.2 OCR Assay

Measuring mitochondrial function is crucial in understanding cellular processes such as activation, proliferation, differentiation, and dysfunction. Directly measuring the oxygen consumption rate (OCR) of cells on the Seahorse Cellular Energy Metabolism Analysis System can provide a comprehensive view of the functional state of mitochondrial metabolism, revealing critical information not available through basal metabolic assays. This assay involves the use of compounds such as oligomycin, which inhibits ATP synthase (complex V) and reduces electron flow through the electron transport chain (ETC), resulting in decreased mitochondrial respiration or OCR upon the first injection. The second injection involves FCCP (carbonylcyanide phenylhydrazone, carbonyl cyanide 4-(trifluoromethoxy) phenylhydrazone), which disrupts the proton gradient and mitochondrial membrane potential, leading to an increase in OCR. This increase in OCR is used to calculate the reserve respiratory capacity, which measures the cell's ability to respond to increased energy demand or stress. The third injection involves a mixture of rotenone, an inhibitor of complex I, and antimycin A, an inhibitor of complex III. This mixture shuts down mitochondrial respiration, enabling the calculation of non-mitochondrial respiratory oxygen consumption driven by extra-mitochondrial processes.

1. Digest the cells with trypsin, resuspend and count them, and inoculate them into Seahorse Energy Metabolizer 24-well plates at a concentration of 15,000–20,000 cells/well, leaving at least one blank well. Once the cells have settled, supplement the medium of each well to 500 μL and place in an incubator at 37 °C and 5% CO_2 until the cells adhere.
2. Open the XF mitochondrial stress assay kit and add 1 mL of Seahorse pH 7.4 buffer to each well of the 24-well plate. Cover the plate, submerge the probe in the buffer, and incubate the entire kit in a CO_2-free incubator at 37 °C overnight for hydration.

3. Once the cells are completely adherent, remove the 24-well plate, discard the medium, and gently wash the cells with fresh medium to detect OCR.
4. Add Oligomycin, FCCP, and rotenone/antimycin A to the plates according to the order of the instructions, then gently move the plate to the machine for calibration, which takes about 20 min.
5. After the cells are processed and corrected, remove the 24-well plate and carefully place the calibrated reaction plate over the cells, then start measuring the OCR values of the cells by sequentially injecting the three inhibitors.
6. Perform data processing.

3.2.3 ECAR Assay

Glucose within the cell is first converted to pyruvate via glycolysis and then further metabolized to lactate in the cytoplasm or to CO_2 and water in the mitochondria. The conversion of glucose to pyruvate and lactate results in a net production of protons that are expelled into the extracellular medium, leading to acidification of the medium surrounding the cell. If the cells predominantly rely on glycolysis, a glycolytic stress analysis assay is recommended to determine the complete function of intracellular glycolysis. The XF instrument directly measures the acidification rate and reports this as the extracellular acidification rate (ECAR). This assay involves the use of agents such as glucose, which reflects the normal glycolytic basal respiration value when both aerobic respiration and glycolysis are present, and 2-DG, a competitive inhibitor of hexokinase that inhibits glucose catabolism and glycolysis. The 2-DG assay measures five key parameters of glycolytic metabolism by detecting the rate of hydrogen ion production in solution via a probe with a fluorescent substance (*see* **Note 2**).

1. The cells undergo the same pre-processing as OCR **steps 1–3**.
2. Glucose, oligomycin, and 2-DG are added to the reaction plate in order according to the instructions, and then gently placed on the machine for calibration, which takes about 20 min.
3. After the cells are treated and calibrated, the 24-well plate is removed and the calibrated reaction plate is carefully covered over the cells to start the detection of ECAR values of the cells. Glucose, oligomycin, and 2-DG are added sequentially.
4. Data processing is conducted.

3.3 Monitoring Mitochondrial Membrane Potential

Maintaining a normal mitochondrial membrane potential is essential for the proper function of oxidative phosphorylation and ATP production. Commercially available fluorescent probes, such as tetramethylrhodamine, methyl ester (TMRM), and JC-1, can be used to sensitively detect changes in mitochondrial membrane potential.

3.3.1 TMRM Assay

TMRM is another cationic fluorescent dye that can permeate the cell membrane with single laser excitation and single fluorescence emission peak. Compared to other fluorescent probes, TMRM accumulates in mitochondria solely in response to changes in membrane potential, is less toxic, and has a low binding rate to organelles. TMRM staining can be optimized for use according to the Abcam commercial kit (ab228569), as described below:

1. Seed cells and grow them to logarithmic growth stage, then treat them with drugs for 12–24 h.
2. Prepare a working solution of TMRM Reagent by diluting it to 100 nM in serum-free media (the actual concentration can be optimized for different cell types).
3. Remove the growth medium from cells by aspiration and wash cells with 1× PBS.
4. Incubate cells in the appropriate volume of TMRM working solution prepared in **step 2**.
5. Incubate cells in the dark in the incubator at 37 °C for 30 min. Aspirate the staining medium.
6. Prepare enough 1× Live Cell Imaging Buffer (LCIS) for all wells based on the same volumes used in **step 2**. Bath cells with the prepared 1X LCIS.
7. View cells using a fluorescence microscope with Ex/Em = 548/573.
8. Analyze fluorescence ratio using Image J software.

3.3.2 JC-1 Assay

JC-1 is a widely used fluorescent probe for detecting changes in mitochondrial membrane potential $\Delta\Psi m$. At high mitochondrial membrane potentials, JC-1 aggregates in the mitochondrial matrix, producing red fluorescence. At low mitochondrial membrane potentials, JC-1 cannot aggregate in the mitochondrial matrix, and instead remains monomeric, producing green fluorescence. To use JC-1 staining, the Abcam commercial kit (ab141387) can be optimized according to the following steps:

1. Seed cells onto a chamber slide and grow them to the logarithmic growth stage. Treat the cells with drugs for 12–24 h.
2. Dilute the JC-1 stock solution (5 mg/mL) to the appropriate working concentration using pre-warmed medium at 37 °C. Mix the solution thoroughly and perform the procedure in a light-proof environment.
3. Discard the cell culture medium and add the appropriate amount of JC-1 working solution. Incubate at 37 °C in a 5% CO_2 incubator for 30 min.
4. Discard the staining solution and wash the cells three times with PBS.

5. Add pre-warmed medium at 37 °C and observe the cells under a laser scanning confocal fluorescence microscope.
6. Analyze the fluorescence ratio using Image J software.

3.4 Monitoring Cell Viability

Ferroptosis inducers such as erastin and RSL3 can significantly inhibit cell viability and the activity of various enzymes, including mitochondrial enzymes. To assess cell viability, commercial reagents can be used to detect related enzyme activity [23].

One commonly used colorimetric detection kit for cell viability is the Cell Counting Kit-8 (CCK-8), which is based on the reduction of WST-8 [(2-(2-methoxy-4-nitrophenyl)-3-(4-nitrophenyl)-5-(2,4-disulfophenyl)-2H-tetrazolium sodium salt)] by dehydrogenases in mitochondria, resulting in the formation of orange-yellow formazan in the presence of electron coupling reagents. The reaction color is positively correlated with the number of living cells. The protocol for using the CCK-8 kit is as follows (*see* **Note 3**):

1. Select cells in the logarithmic growth phase and prepare cell suspensions.
2. Inoculate 100 μL of cell suspensions in 96-well plates at a density determined by the cell growth rate. Set up 6 replicate wells.
3. Pre-culture the plates in an incubator (37 °C, 5% CO_2) for 24 h to allow the cells to reach the exponential phase.
4. Discard the cell culture medium.
5. Prepare a working solution of CCK-8 (final concentration of 10%) in serum-free cell culture medium (Medchemexpress; cat: HY-K0301).
6. Incubate the plate in the incubator for 0.5–4 h (the best reaction time for CCK-8 depends on the optimal time for specific color development; the best OD value is around 1.0).
7. Detect the absorbance (OD value) at a wavelength of 450 nm using a microplate reader. Repeat the experiment three times and take the average value of the results as the final experimental result.
8. Add the same volume of CCK-8 to cell-free complete medium and incubate for the same duration, and measure the absorbance at 450 nm together with the experimental group as the blank group.
9. Analyze the data obtained.

3.5 Monitoring Mitochondrial ROS

Lipid peroxidation is a characteristic feature of ferroptosis, and GPX4 is a major antioxidant regulator. Inactivation of GPX4 can unleash potent ferroptotic effects in many cell lines [24–26]. However, despite this, no significant mitochondrial lipid peroxidation was detected using the C11-BODIPY probe. Paradoxically, another study showed that the mitochondria-targeted antioxidant Mito-TEMPO can block doxorubicin-induced ferroptosis in mouse hearts, which provides strong in vivo evidence that mitochondria are linked to ferroptosis [27]. Thus, the detection of the lipid peroxidation status of mitochondria during ferroptosis is crucial for understanding the context-dependent role of mitochondria.

The MitoSOX superoxide indicator is a fluorescent dye that specifically targets mitochondria in living cells. MitoSOX is oxidized by mitochondrial superoxide to produce bright fluorescence. Therefore, it is recommended to use MitoSOX to detect mitochondrial ROS, as described below:

1. Seed cells onto chamber slides, allow them to grow to the logarithmic growth phase, and treat them with drugs for 12–24 h.
2. Mix 50 μg of MitoSOX Red Mitochondrial Superoxide Indicator (Thermo Fisher) with 13 μL of dimethyl sulfoxide (DMSO). Prepare a 5 mM stock solution.
3. Discard the cell culture medium.
4. Dilute the stock solution 1000 times with serum-free DMEM medium to a final concentration of 5 μM. (Adjust the concentration as needed, but not exceeding 5 μM.)
5. Add 1 mL of the probe working solution to cover the cells to be analyzed and incubate at 37 °C for 30 min under low light.
6. Wash the cells 3 times with preheated serum-free medium at 37 °C.
7. Add fresh medium to incubate the cells.
8. Detect the fluorescence using a fluorescence confocal microscope. (Maximum absorption/emission wavelength ~396/610 nm).

3.6 Monitoring Mitophagy

Mitophagy serves as a mechanism for mitochondrial quality control [28–31]. Activated mitophagy induces ROS production, thereby increasing the sensitivity of ferroptosis. Additionally, zalcitabine-induced mtDNA stress also promotes autophagy-dependent ferroptosis [10]. Therefore, it is essential to monitor the correlation between mitochondria and autophagy. To monitor mitochondrial autophagy levels during ferroptosis, it is reasonable to use a mitochondrial labeling probe with a microtubule-associated protein 1 light chain 3 (MAP1LC3; also known as LC3) co-localization assay, as described below (*see* **Note 4**).

1. Calculate the mRFP-GFP-LC3B fusion protein adenovirus MOI. MOI (Multiplicity of infection): The ratio of the number of viruses to the number of cells when a virus infects a cell.

2. Determine the required pfu: cell number × MOI. Pfu (Plaque forming units): the number of biologically active virus particles.
3. The day before infection, inoculate the target cells in a 6-well plate with approximately 5 × 10^5 cells/well (the exact number of inoculations depends on the cell size and cell growth rate). Add 2 mL DMEM medium with 10% FBS to each well so that the cell density reaches approximately 50% the next day when the virus is infected.
4. Calculate the amount of virus required according to the MOI of 2, 5, 10, 20, and 40.
5. Add 1.2 mL of fresh culture medium to each well, then add virus solution with a specific MOI value to each well.
6. After 24 h, discard the culture medium containing the virus and add 2 mL of fresh complete culture medium to each well. Incubate for another 24 h.
7. Treat the infected cells with the required drugs.
8. Discard the medium and add MitoTracker™ Red (Thermofisher; cat: M7512) at a concentration of 1 μM, then incubate for 30 min.
9. Wash the cells with PBS three times.
10. Detect the cells using laser confocal fluorescence microscopy.

4 Notes

1. Western blotting is a complex multi-step experiment and the factors affecting the results may be diverse. Therefore, if the final protein imaging results are poor, optimization should be gradually excluded according to the actual situation.
2. There are six parameters used to assay ECAR: (1) Basic Respiration: The energy used to meet cellular ATP requirements and oxygen consumption for mitochondrial proton leakage. (2) Proton Leakage: The remaining uncoupled basal respiration generated by the combination of ATP and proton leakage. Proton leakage can be an indication of mitochondrial damage. (3) Maximal Respiration: The maximum oxygen consumption rate obtained after adding the uncoupling agent FCCP. (4) Reserve Capacity: Indicates the cell's ability to cope with energy demand and the difference between cellular respiration and its theoretical maximum. (5) Non-mitochondrial Respiration: The oxygen depletion that remains after the addition of rotenone and oligomycin A, which is due to the continued consumption of oxygen by a portion of the enzymes in the cells. (6) Glycolysis Reserves: Indicates the cell's ability to

respond to energy demand and how close the cellular glycolytic function is to the theoretical maximum.

3. To reduce measurement errors, the outermost ring of the plate is left empty, and 100 μL of PBS is added. During cell inoculation, it is important to mix the cell suspension thoroughly to prevent settling and ensure consistent cell numbers in each well.
4. An infection efficiency of 20–70% is optimal for detecting autophagy, as it provides sufficient levels of infection without causing significant cytotoxicity or interfering with the detection process. Higher infection efficiencies can easily lead to cytotoxic effects and interfere with the accurate detection of autophagy.

Acknowledgments

Research by D.T. and R.K. was supported by grants from the National Institutes of Health (R01CA160417, R01CA229275, and R01CA211070).

References

1. Andrews NC (1999) Disorders of iron metabolism. N Engl J Med 341:1986–1995
2. Tang D, Chen X, Kang R et al (2021) Ferroptosis: molecular mechanisms and health implications. Cell Res 31:107–125
3. Chen X, Li J, Kang R et al (2021) Ferroptosis: machinery and regulation. Autophagy 17: 2054–2081
4. Chen X, Huang J, Yu C et al (2022) A noncanonical function of EIF4E limits ALDH1B1 activity and increases susceptibility to ferroptosis. Nat Commun 13:6318
5. Zhang R, Kang R, Tang D (2023) Ferroptosis in gastrointestinal cancer: from mechanisms to implications. Cancer Lett 561:216147
6. Chen X, Kang R, Kroemer G et al (2021) Organelle-specific regulation of ferroptosis. Cell Death Differ 28:2843–2856
7. Bock FJ, Tait SWG (2020) Mitochondria as multifaceted regulators of cell death. Nat Rev Mol Cell Biol 21:85–100
8. Xie Y, Hou W, Song X et al (2016) Ferroptosis: process and function. Cell Death Differ 23: 369–379
9. Li Y, Wang X, Huang Z et al (2021) CISD3 inhibition drives cystine-deprivation induced ferroptosis. Cell Death Dis 12:839
10. Li C, Zhang Y, Liu J et al (2021) Mitochondrial DNA stress triggers autophagy-dependent ferroptotic death. Autophagy 17: 948–960
11. Gao M, Yi J, Zhu J et al (2019) Role of mitochondria in ferroptosis. Mol Cell 73(354–63): e3
12. Yuan H, Li X, Zhang X et al (2016) CISD1 inhibits ferroptosis by protection against mitochondrial lipid peroxidation. Biochem Biophys Res Commun 478:838–844
13. Liu Y, Wang Y, Lin Z et al (2023) SLC25A22 as a key mitochondrial transporter against ferroptosis by producing GSH and MUFAs. Antioxid Redox Signal 39(1–3):166–185
14. Dixon SJ, Lemberg KM, Lamprecht MR et al (2012) Ferroptosis: an iron-dependent form of nonapoptotic cell death. Cell 149:1060–1072
15. Mao C, Liu X, Zhang Y et al (2021) DHODH-mediated ferroptosis defence is a targetable vulnerability in cancer. Nature 593:586–590
16. Yuan H, Li X, Zhang X et al (2016) Identification of ACSL4 as a biomarker and contributor of ferroptosis. Biochem Biophys Res Commun 478:1338–1343
17. Yang WS, SriRamaratnam R, Welsch ME et al (2014) Regulation of ferroptotic cancer cell death by GPX4. Cell 156:317–331

18. Feng H, Schorpp K, Jin J et al (2020) Transferrin receptor is a specific ferroptosis marker. Cell Rep 30(3411–23):e7
19. Liu J, Zhu S, Zeng L et al (2021) DCN released from ferroptotic cells ignites AGER-dependent immune responses. Autophagy 18: 2036–2049
20. Lee H, Zandkarimi F, Zhang Y et al (2020) Energy-stress-mediated AMPK activation inhibits ferroptosis. Nat Cell Biol 22:225–234
21. Schmitt M, Ceteci F, Gupta J et al (2022) Colon tumour cell death causes mTOR dependence by paracrine P2X4 stimulation. Nature 612:347–353
22. Zhang Z, Deng W, Kang R et al (2016) Plumbagin protects mice from lethal sepsis by modulating immunometabolism upstream of PKM2. Mol Med 22:162–172
23. Chen X, Song X, Li J et al (2023) Identification of HPCAL1 as a specific autophagy receptor involved in ferroptosis. Autophagy 19:54–74
24. Dai E, Han L, Liu J et al (2020) Ferroptotic damage promotes pancreatic tumorigenesis through a TMEM173/STING-dependent DNA sensor pathway. Nat Commun 11:6339
25. Li J, Liu J, Xu Y et al (2021) Tumor heterogeneity in autophagy-dependent ferroptosis. Autophagy 17:3361–3374
26. Zhu S, Zhang Q, Sun X et al (2017) HSPA5 regulates ferroptotic cell death in cancer cells. Cancer Res 77:2064–2077
27. Fang X, Wang H, Han D et al (2019) Ferroptosis as a target for protection against cardiomyopathy. Proc Natl Acad Sci U S A 116: 2672–2680
28. Chen F, Cai X, Kang R et al (2023) Autophagy-dependent ferroptosis in cancer. Antioxid Redox Signal 39(1–3):79–101
29. Liu J, Kuang F, Kroemer G et al (2020) Autophagy-dependent ferroptosis: machinery and regulation. Cell Chem Biol 27:420–435
30. Xie Y, Liu J, Kang R et al (2020) Mitophagy receptors in tumor biology. Front Cell Dev Biol 8:594203
31. Xie Y, Liu J, Kang R et al (2021) Mitophagy in pancreatic cancer. Front Oncol 11:616079

Chapter 11

Generation of Organoids and Analysis of Ferroptosis in Organoids

Wenxin Li, Yujie Su, Jingyi Guo, Mengfei Wang, and Xingguo Liu

Abstract

Ferroptosis is a unique form of iron-dependent cell death induced by lipid peroxidation and subsequent plasma membrane rupture, which sets it apart from other types of regulated cell death. Ferroptosis has been linked to a diverse range of biological processes, such as aging, immunity, and cancer. Organoids, on the other hand, are three-dimensional (3D) miniaturized model systems of different organs in vitro cultures, which have gained widespread interest for modeling tissue development and disease, drug screening, and cell therapy. Organoids offer tremendous potential for improving our understanding of human diseases, particularly in the search for the field of ferroptosis in pathological processes of organs. Furthermore, cancer organoids are utilized to investigate molecular mechanisms and drug screening in vitro due to the anti-tumor effect of ferroptosis. Currently, the development of liver organoids has reached a relatively mature stage. Here, we present the protocols for the generation of liver organoids and liver cancer organoids, along with the methods for detecting ferroptosis in organoids.

Key words Ferroptosis, Organoids generation, Cancer organoids

1 Introduction

In 2012, Dixon and colleagues introduced the term ferroptosis, which refers to a unique iron-dependent form of cell death triggered by the accumulation of lipid peroxidation [1]. Ferroptosis is distinct from other forms of cell death, including apoptosis, necroptosis, and pyroptosis, in terms of cell morphology and function. It has been implicated in various diseases, including liver diseases, heart diseases, neurodegenerative diseases, and hematological diseases [2]. Over the past decade, an increasing number of disease models have been established where ferroptosis plays a crucial role in organ damage [3, 4]. With the development of 3D

Authors Wenxin Li and Yujie Su have equally contributed to this chapter.

Guido Kroemer and Daolin Tang (eds.), *Ferroptosis: Methods and Protocols*, Methods in Molecular Biology, vol. 2712,
https://doi.org/10.1007/978-1-0716-3433-2_11,

culture systems, organoids have become a powerful tool for disease modeling.

Since the first intestine organoids were induced in vitro by Lgr^{5+} intestinal stem cells [5], organoids constitute an appealing approach to model disease development. Two-dimensional (2D) cultured cells lack essential information in cell-cell communication, while organoids mimic complex tissue architecture and organ function, providing a closer physiological environment [6]. Organoid technology has been applied to multiple tissues, including the liver [7], the brain [8], the intestine [9], and the kidney [10]. Multiple liver cancer organoids models have also been developed, such as hepatocellular carcinoma (HCC), cholangiocarcinoma (CC), and combined HCC/CC (CHC) [11]. A Liver organoid-based Toxicity screen (LoT) was developed, which has demonstrated high predictive values for marketed drugs based on viability and cholestatic effects, and successfully predicted CYP2C9*2 as a genomic predisposition for Bosentan-induced cholestasis [12].

Currently, several methods based on different targets have been developed for detecting ferroptosis at both cellular and organoid levels. Specific morphological changes associated with ferroptosis, such as plasma membrane rupture, mitochondrial volume decrease, cristae disappearance, and an increase in ultrastructures related to autophagy, can be observed using light or electron microscopy [13, 14]. The biomarkers of ferroptosis can be identified from iron accumulation, lipid peroxidation, and glutathione (GSH) reduction pathways [15]. At the organoid level, there are two main methods to detect cell death: cell mortality and cell viability. Ferroptosis, in particular, can be monitored by measuring the levels of ferrous ion (Fe^{2+}), glutathione (GSH), and malondialdehyde (MDA) in organoids [16]. Therefore, it is necessary to develop more specific methods for detecting ferroptosis at the organoid level.

Organoids have become a popular tool in studying various human diseases, including the liver. As an example, we will explore the methods for generating liver organoids and detecting ferroptosis in these organoids.

2 Materials

The current protocol involves the use of the following materials:

2.1 Generation of Liver Organoids from iPSCs

1. Biological materials:human fibroblast cells (*see* **Note 1**).
2. Human fibroblast cells medium: Minimum essential medium (MEM) supplemented with 10% fetal bovine serum (FBS), 1%

non-essential amino acids (NEAA), 1% pyruvate sodium and 200 mM uridine.

3. 293 T cells.
4. Dulbecco's modified eagle medium (DMEM).
5. Penicillin-Streptomycin.
6. Lipofectamine 3000.
7. Defined FBS (DFBS).
8. Beta-mercaptoethanol (β-ME).
9. Basic fibroblast growth factor (bFGF).
10. Feeder cells.
11. Vitamin C.
12. DMEM/F12.
13. Knockout serum replacement (KSR).
14. Human embryonic stem cell culture medium (mTeSR1 medium).
15. Definitive endoderm differentiation medium: Roswell Park Memorial Institute (RPMI) 1640 supplemented with 1:50 B27 without insulin, 1% GlutaMax, 1% Penicillin-Streptomycin and 100 ng/mL human activin A.
16. Hepatic specification medium: RPMI 1640 medium supplemented with 1:50 B27 with insulin, 1% GlutaMax, 1% Penicillin-Streptomycin, 20 ng/mL bone morphogenetic protein 2 (BMP2) and 30 ng/mL fibroblast growth factor 4 (FGF4).
17. Hepatic expansion medium: RPMI 1640 medium supplemented with 1:50 B27 with insulin, 1% GlutaMax, 1% Penicillin-Streptomycin, 20 ng/mL hepatocyte growth factor (HGF) and keratinocyte growth factor (KGF).
18. Hepatic maturation medium: Hepatocyte Culture Medium supplemented with SingleQuots and 20 ng/mL Oncostatin-M (OSM).
19. Hepatic medium.
20. N-2-hydroxyethylpiperazine-N-ethane-sulphonicacid (HEPES).
21. R-spondin-1 (RSPO1).
22. N-acetylcysteine (NAC).
23. Epidermal growth factor (EGF).
24. Gastrin.
25. Laduviglusib (CHIR-99021).
26. Transforming growth factor alpha (TGFα).
27. Paraformaldehyde (PFA).

28. Sucrose.
29. PBS.
30. Diamidino-phenyl-indole (DAPI).
31. Rho kinase (ROCK) inhibitor Y-27632 dihydrochloride.
32. Cell culture incubator (37 °C, 5% CO_2).
33. Cryostat microtome.
34. Enzyme linked immunosorbent (ELISA) Assay Kit.
35. Microplate reader.
36. P450-Glo Assay Kit.

2.2 Generation of Human Cancer Organoids

1. Biological materials:human liver cancer specimen (*see* **Note 1**).
2. Rho kinase (ROCK) inhibitor Y-27632 dihydrochloride.
3. Collagenase IV.
4. Dnase.
5. Cultrex Reduced Growth Factor Basement Membrane Extract, Type 2, Pathclear (RGF BME).
6. Human liver cancer organoid isolation medium: Advanced DMEM/F12 supplemented with 1% Penicillin/Streptomycin, 1% Glutamax, 10 mM HEPES, 1:50 B27 supplement (without Vitamin A), 1:100 N2, 1.25 mM NAC, 10 mM nicotinamide, 10 nM recombinant human [Leu15]-Gastrin I, 50 ng/mL recombinant human EGF, 100 ng/mL recombinant human FGF10, 25 ng/mL recombinant human HGF, 10 μM Forskolin, 5 μM A83–01, 10 μM Y27632 and 3 nM Dexamethasone and pre-warmed medium in 37 °C water bath.
7. Human liver cancer organoid expansion medium: Advanced DMEM/F12 supplemented with 1% Penicillin/Streptomycin, 1% Glutamax, 10 mM HEPES, 1:50 B27 (without Vitamin A), 1:100 N2, 1.25 mM NAC, 10% (vol/vol) RSPO1 conditioned medium, 10 mM nicotinamide, 10 nM recombinant human [Leu15]-Gastrin I, 50 ng/mL recombinant human EGF, 100 ng/mL recombinant human FGF10, 25 ng/mL recombinant human HGF, 10 μM Forskolin and 5 μM A83–01 and pre-warmed in a 37 °C water bath.

2.3 Monitoring Ferroptosis of Organoids

1. Detection of cell death: S7020-SYTOX® Green nucleic acid stain *5 mM solution in DMSO.
2. Detection of cell viability: CellTiter-Glo® 3D Cell Viability Assay.
3. Detection of lipid peroxides: Lipid Peroxidation (MDA) Assay kit.

4. Detection of intracellular Fe^{2+} content: Iron Assay kit.
5. Detection of GSH: GSH Assay Kit.

3 Methods

3.1 Generation of Liver Organoids from iPSCs

The liver is composed of several cell types, including hepatocytes, bile duct cells, Kupffer cells, and stellate cells. Organoids technology can replicate liver features for drug screening and toxicity testing [12, 17, 18]. Liver organoids can be derived from adult stem cells (ASCs), embryonic stem cells (ESCs), and induced pluripotent stem cells (iPSCs) [17]. Due to ethical and resource constraints of ASCs and ESCs, iPSCs have been the most promising source for generating liver organoids [19].

For organoid generation, iPSCs are sequentially cultured in a specific medium containing external growth factors to induce specific cell types. Different materials are used to construct a 3D microenvironment and replicate the native extracellular matrix (ECM) of the tissue to support cell adhesion and growth [20]. Matrigel, a natural ECM derived from Engelbreth-Holm-Swarm (EHS) mouse sarcoma [21], has been the most widely used matrix for organoid generation. However, matrigel contains unknown cytokines and potential uncertainty for its application, particularly in transplant therapy. Chemically defined synthetic scaffolds, such as hydrogels, have been proposed as an alternative material for organoid generation [22, 23].

Different markers are expressed at different stages of liver differentiation. Quantitative polymerase chain reaction (qPCR) and immunofluorescence are used to detect the definitive endoderm marker SOX17, hepatoblasts marker α-fetoprotein (AFP), and mature hepatocytes marker albumin (ALB). QPCR is also used to detect the highly expressed hepatocytes specific genes, such as liver functional protein genes (*ALB*, *AFP*, *AAT*), transport protein genes (*NTCP*, *MRP3*), liver-specific transcriptional factors (*HNF3α*, *HNF3b*), and metabolism enzyme genes (*CYP3A4*, *CYP2C9*, *CYP2C19*) [17, 24]. The induced hepatocyte organoids (iHep-Orgs) are then analyzed by the marker of ALB, the accumulation of glycogen, the mRNA levels of the hepatocyte genes (cytochrome P450 activity, glycogen/lipid metabolism, and urea cycle), and so on to assess the function of iHep-Orgs in vitro [7].

We present a detailed methodology for generating liver organoids from iPSCs.

3.1.1 iPSCs Generation

1. Prepare the human fibroblast cells medium.
2. Human fibroblast cells are cultured in the human fibroblast cells medium in a cell culture incubator (37 °C, 5% CO_2), and the cells are passed every week.

3. Package of retroviruses encoding human *SOX2*, *KLF4*, *OCT4*, and *c-MYC*.
   (a) Construct the retroviruses vector expressed human *SOX2*, *KLF4*, *OCT4*, and *c-MYC*.
   (b) 293 T cells are seeded at 70% to 90% confluency in high glucose DMEM medium supplement with 10% FBS and 1% Penicillin-Streptomycin.
   (c) Gently mix plasmids with transfection reagents (such as Lipofectamine 3000) and incubate for 10–15 min at room temperature, then transfer to the medium.
   (d) Incubate the cells overnight and replace with fresh medium.
   (e) Collect the supernatant containing the virus after 4 days.
   (f) Filter the virus with a 0.45 um filter membrane and add polybrene to 8 ug/mL.
4. Infect fibroblast cells with virus encoding human *SOX2*, *KLF4*, *OCT4*, and *c-MYC* twice.
5. Culture fibroblast cells in high glucose DMEM supplement with 20% hESC-DFBS, 1% NEAA, 1% GlutaMax, 0.1% β-ME, and 0.1% bFGF.
6. After 6 days of culturing, the cells are seeded on 10 cm dishes coated with feeder cells and Vitamin C (50 μg/mL) and VPA (1 mM) are added for 7 days to enhance the reprogramming process.
7. Switch the medium with KSR-based medium composed of DMEM/F12 supplement with 20% KSR, 1% NEAA, 1% GlutaMax, 0.1% β-ME, and 0.1% bFGF.
8. Human ESC-like colonies are generated after approximately 20–25 days of infection and manually picked between days 25 and 30.
9. The iPSCs colonies are then maintained in mTeSR1 medium.
10. The expression of pluripotent genes such as *OCT4*, *SOX2*, *NANOG*, and *REX1* are detected by qPCR.

3.1.2 Hepatocytes Generation

1. IPSCs are cultured on Matrigel-coated dishes under PSC medium for 3 days, followed by a switch to the definitive endoderm differentiation medium and incubated for 3 days. The expression of *FOXA2*, *SOX17*, and *GATA4* could then be detected by qPCR or immunocytochemistry using human antibodies.
2. Prepare the hepatic specification medium, the hepatic expansion medium, and the hepatic maturation medium.

3. Then, the medium is exchanged with the hepatic specification medium for 4 days. The expression of *FOXA2* and *HNF4a* could then be detected by qPCR or immunocytochemistry.
4. The medium is exchanged with the hepatic expansion medium for 6 days. The expression of *FOXA2*, *HNF4a*, and *AFP* could then be detected by qPCR or immunocytochemistry using human antibodies.
5. For hepatic maturation, the cells are cultured by the hepatic maturation medium for 8 days. Albumin and the hepatocyte markers *ALB*, *HNF4a*, *CYP1A2*, and *CYP1A11* can be detected by qPCR or immunocytochemistry using human antibodies.
6. Approximately 9 to 12 days after hepatic maturation, a 3D spherical structure can be observed over the 2D monolayers of mature hepatocytes.

3.1.3 Liver Organoids Generation

1. To solidify cysts and form a dome, Matrigel should be used. The hepatic medium should be supplemented with 10 nM HEPES, 1% GlutaMax and 1% Penicillin-Streptomycin plus 15% RSPO1 conditioned medium, 1:50 B27 (minus vitamin A), 50 ng/mL EGF, 1.25 mM NAC, 10 nM gastrin, 3 mM CHIR99021, 50 ng/mL HGF, 100 ng/mL FGF7, 100 ng/mL FGF10, 2 mM A83–01, 10 mM Nicotinamide, 10 mM Y-27632 and 20 ng/mL TGFα.
2. Functional analysis of liver organoids.
   (a) To access glycogen storage, organoids are fixed with 4% PFA, cryo-protected in 30% sucrose, and frozen in optimal-cutting-temperature compound. The frozen sections are then sliced to a thickness of 10 μM at −20 °C using a cryostat microtome. The sectioned samples are then stained with periodic acid-Schiff according to the manufacturer's instructions.
   (b) Low-density lipoprotein (LDL) uptake is detected using DiI-Ac-LDL. The organoids are treated with 20–50 μg/mL for 4 h at 37 °C, then the organoids are washed, fixed, and incubated with DAPI and imaged by immunofluorescence.
   (c) For album secretion detection, the supernatant of the organoids is collected and determined using Bethyl ELISA Kit.
   (d) Human A1-Antitrypsin (ATT) and AFP are detected using ELISA kits. The absorbance is detected by a Spectra Max M3 microplate reader, and the data are normalized by the cell number.

(e) For Cytochrome P450 (CYP450) activity, CYP450 enzymes are induced by treatment with each inducer (CYP3A4: 20 μM rifampicin, 100 μM acetaminophen, and 10 μM nifedipine and CYP1A2: 150 μM omeprazole) for 48 h. CYP enzyme activities are measured after 3 h of incubation with each subtype-specific substrate using a P450-Glo Assay Kit.

3. Replace the medium every 3 days.
4. Passage organoids every 7 days. Wash organoids with cold PBS to remove the Matrigel and split them into small pieces using a surgical blade under a dissecting microscope. The passaged organoids should then be resuspended in Matrigel in a ratio of 1:3 to 1:10.
5. Cryopreserve organoids. Mix passaged organoids with mFreSR™ and freeze organoids.

3.2 Generation of Human Cancer Organoids

The research on ferroptosis in cancers has rapidly accumulated in recent years; however, there are still many challenges, particularly in the construction of cancer models. Currently, the most commonly used cancer models are patient-derived 2D cancer cell line (PDC), patient-derived cancer xenotransplantation model (PDTX), and 3D organoids. Cancer cell lines may undergo genetic changes during passage [25], and due to the lack of intercellular connectivity and extracellular matrix, they are more sensitive to drugs, which can lead to false positive data for drug screening [26]. The model of human cancer xenotransplantation has some disadvantages such as low success rate and long cycle, and PDTX may also experience mouse-specific cancer evolution [27]. 3D organoids culture technology has been used to develop new and more physiological cancer models, which demonstrates strong advantages due to its high speed, high drug screening flux, and strong clinical correlation [28].

Cancer organoids such as liver cancer, pancreatic cancer, colorectal cancer, and prostate cancer have been successfully obtained. Taking liver cancer organoids as an example, we discuss and summarize the methods of cultivating liver cancer organoids. The primary liver cancer is classified into HCC, CC, and CHC subtypes. Cancer organoids can be obtained from surgical resection and needle biopsies of cancer tissue [11, 29], as well as from iPSCs [30]. However, the efficiency of generating iPSC-based cancer organoids from patients may depend on the type of cancer and the presence or absence of specific carcinogenic mutations, thus cancer organoids derived from cancer tissues are more widely used [25].

Cancer tissues obtained from surgical resection and needle biopsy are digested at 37 °C. Cells are collected and mixed with RGF BME, and 3000–10,000 cells are inoculated per well in a

24-multiwell plate. When RGF BME is solidified, the separation medium is added and changed every 2–3 days. After 2–3 weeks of culture, the cultures are mechanically separated into small fragments and transferred to the human liver-derived organoid expansion medium which is changed every 2–3 days. Then, the cancer-derived organoids are passaged at the proportion of 1:4–1:8 every 7–10 days for at least 6 months.

3.2.1 Generation of Hepatocellular Carcinoma-Derived Organoids

1. Obtain fresh HCC, CCC, and CHC tissue by surgical resection or obtain multiple biopsy samples by ultrasound (US) guided biopsy.
2. Cancer tissues and normal tissues should be placed in 45 mL cold Advanced DMEM/F12 medium. (Optional) Add 10 μM Y-27632 to Advanced DMEM/F12 medium to increase cell survival (*see* **Note 2**).
3. Carefully assess obtained tissue pieces and remove fat, muscle tissue, and non-epithelial components as much as possible using surgical scissors or scalpels under a dissection microscope.
4. Mince tissues into small fragments of 1 to 3 mm^3 in a 10-cm cell culture dish using surgical scissors or scalpels.
5. Digest tissues with 2.5 mg/mL collagenase IV and 0.1 mg/mL DNase at 37 °C. Digest cancer-derived tissue sections for 2–5 h even overnight. Digest puncture samples for 10 min. When a mixture of cell clusters composed of 2 to 10 cells is observed, the digestion is complete.
6. Add 10 mL of Advanced DMEM/F12 medium and centrifuge at 200 g for 5 min at 4 °C to wash the sample once.
7. Resuspend the pellet in 10 mL of Advanced DMEM/F12 medium and filter with a 100 μM cell strainer. Centrifuge the suspended and filtered cells at 200 g for 5 min at 4 °C.
8. Aspirate the supernatant and resuspend the pellet in 70% (vol/-vol) RGF BME. Keep the RGF BME on ice to prevent it from solidifying. Approximately 10,000 cells should be contained in 40 μL of RGF BME.
9. Seed the RGF BME-covered cells in pre-warmed 24-well culture plates with 2000 to 5000 cells per well.
10. Place the plate into an incubator for 20 to 30 min to allow the RGF BME to solidify.
11. Prepare the human liver cancer organoid isolation medium.
12. Once the RGF BME droplets have solidified, carefully add 500 μL of medium to each well.

13. Change the medium every 2 to 3 days by carefully aspirating the medium from the well and replacing it with fresh, pre-warmed medium (*see* **Note 3**).
14. After 2–3 weeks, pipette the organoids to disrupt the RGF BME and transfer the organoids suspension to a 15 mL conical tube.
15. Fully remove the RGF BME, add 10 mL of ice-cold Advanced DMEM/F12 medium and centrifuge the organoids at 200 g for 5 min at 4 °C.
16. Aspirate the supernatant and split the organoids through mechanical disruption. Resuspend the pellet in 3 mL of Advanced DMEM/F12 medium. Use a P1000 filter pipette tip accompanied by a (non-filtered) P10 tip to pipette the organoids suspension 30 times.
17. After completely shearing, wash the organoids with 10 mL of Advanced DMEM/F12 medium.
18. Resuspend organoid pellets in 70% (vol/vol) RGF BME, and pipette the organoids into small droplets on a pre-warmed plate as described in **steps 8–10**. After completely solidifying, invert the plate and culture it in an incubator (37 °C, 5% CO_2) for 30 to 60 min.
19. Prepare the human liver cancer organoid expansion medium.
20. Carefully add 500 μL medium to each well.
21. Organoids are cultured in an incubator (37 °C, 5% CO_2). The medium is changed every 2 or 3 days by carefully aspirating the medium from the wells and replacing it with fresh, pre-warmed medium.

3.3 Monitoring Ferroptosis of Organoids

Ferroptosis is a non-apoptotic and oxidative damage-related regulated cell death (RCD), which is mainly driven by iron accumulation, lipid peroxidation, and subsequent plasma membrane rupture [31]. Organoids can be used in ferroptosis research because they can well reflect the physiological and pathological function of organs. Therefore, how to detect ferroptosis in organoids is of great significance. At present, the methods for detecting cell death of organoids include PI staining, lactate dehydrogenase (LDH), SYTOX Green [16, 32, 33], and the methods for detecting cell viability of organoids include CCK-8 and Cell-Titer Glo [16, 34–36]. To detect biochemical signatures, Iron Assay Kit, MDA Assay Kit, and GSH Assay Kit are used to measure the contents of Fe^{2+}, lipid peroxides, and GSH in organoids to monitor ferroptosis [16]. To detect the expression of key genes, the mRNA levels of glutathione peroxidase 4 (*GPX4*), ferritin heavy chain (*FTH1*), and cyclooxygenase 2 (*COX2*) are detected by qPCR, and the protein levels of GPX4, FTH1, and COX2 are detected by immunofluorescence [16].

SYTOX Green is a high-affinity nucleic acid dye used to detect cell death, which can easily penetrate dead cells with damaged cell membranes without penetrating living cells. After briefly incubating with SYTOX Green, dead cells emit bright green fluorescence. Compared with PI staining, SYTOX Green has stronger binding affinity to nucleic acid and simpler operation, which makes it more advantageous for its application in the study of ferroptosis.

ATP participates in a variety of enzymatic reactions in organisms, which is an index of metabolism of living cells, and its content directly reflects the number and state of cells. The Cell-Titer Glo photoluminescence method is used to measure the cell viability of 3D cultured cells by quantifying ATP, and has the advantages of fast response, high sensitivity, and stable luminescence signal.

Excessive lipid peroxidation is a sign of ferroptosis [37]. Lipid peroxidation is the process of ROS oxidation of biofilm after the enhancement of oxygen stress. Lipid peroxides include MDA and 4-hydroxynonenal (4-HNE). At present, the MDA Assay Kit can be used to detect the level of MDA in organoids [16]. MDA in the samples reacts with thiobarbituric acid (TBA) to form MDA-TBA adducts, which can be easily quantified by colorimetry (OD = 532 nm) or fluorescence method (Ex/Em = 532 nm).

Excessive Fe^{2+} can not only increase the production of ROS through the Fenton reaction, but also activate iron-containing enzymes such as lipoxygenases (LOXs) to promote the formation of lipid peroxides and induce ferroptosis [38]. Therefore, monitoring the intracellular Fe^{2+} content is crucial for the assessment of ferroptosis. Under acidic conditions (pH less than 5.5), iron no longer has a binding affinity for the carrier protein and will be dissociated and released into the solution as iron/ferric ion. Free Fe^{2+} can react with iron probe to produce a stable-colored complex with absorbance at OD=593 nm.

The selenium-containing GPX4 is currently recognized as a central repressor of ferroptosis [39]. The exhaustion of GSH directly impacts GPX4 activity and stability, thus predisposing cells to succumb to ferroptosis [40]. The GSH Assay Kit is based on a specific enzymatic cycling method in the presence of GSH and a fluorophore. The reduction of the fluorophore produces a stable fluorescent product, the fluorescence of which is directly proportional to the amount of GSH in the sample and can be monitored kinetically (Ex/Em = 535/587 nm).

3.3.1 SYTOX Green Assay

A SYTOX Green assay protocol (based on typical organoids culture) is described in detail as follows (Guo et al., 2021):

1. Wash the organoids 1–3 times in a phosphate-free buffer as needed.

2. Prepare the SYTOX Green staining solution by diluting the stock solution 1:30,000 (167 nM) in a phosphate-free buffer.
3. Add sufficient staining solution to cover the organoids.
4. Incubate for 15–30 min, protected from light.
5. Remove the staining solution.
6. Wash the organoids 2–3 times in a phosphate-free buffer.
7. Observe the staining by fluorescence microscope.
8. Use Image J software to quantify the ratio of dead cells (SYTOX Green-positive) to total cells.

3.3.2 Cell-Titer Glo Assay

A Cell-Titer Glo assay protocol (based on typical organoids culture) is described in detail as follows (Ouyang et al., 2022; Zhang et al., 2022; Zou et al., 2022):

1. Thaw the CellTiter-Glo® 3D Reagent at 4 °C overnight.
2. Equilibrate the CellTiter-Glo® 3D Reagent to room temperature by placing the reagent in a 22 °C-water bath prior to use for 30 min approximately.
3. Mix gently by inverting the contents to obtain a homogeneous solution (*see* **Note 4**).
4. Prepare ATP standard curve dilution.
5. Add a volume of CellTiter-Glo® 3D Reagent equal to the volume of cell culture medium and ATP standard present in each well (*see* **Note 5**).
6. Mix the contents vigorously for 5 min to induce cell lysis (*see* **Note 6**).
7. Allow the plate to incubate at room temperature for an additional 25 min to stabilize the luminescent signal.
8. Record the luminescence (*see* **Note** 7).
9. Compare luminescence of samples to luminescence of standards to determine ATP detected by the CellTiter-Glo® 3D Reagent in samples.

3.3.3 MDA Assay

A MDA assay protocol (based on typical organoids culture) is described in detail as follows [16]:

1. Prepare the necessary reagents in advance.
2. Prepare colorimetric or fluorometric assay standard curve dilution.
3. Harvest the number of organoids needed for each assay (initial recommendation = 1–10 mg) and repeat 3 times for each sample.
4. Wash organoids in cold PBS.

5. Mix 300 μL MDA lysis buffer with 3 μL butylated hydroxytoluene (BHT) (100X) to prepare lysis solution.
6. Homogenize organoids in 303 μL Lysis Solution (Buffer + BHT) with a Dounce homogenizer sitting on ice, with 10–15 passes.
7. Centrifuge at 13,000 g for 10 min to remove insoluble material. Collect supernatant to a new tube and place it on ice.
8. Add 600 μL of TBA reagent into each well containing 200 μL standards and 200 μL samples to generate MDA-TBA adduct.
9. Incubate at 95 °C for 60 min. Cool to room temperature in an ice bath for 10 min (*see* **Note 8**).
10. Take 200 μL of supernatant containing MDA-TBA adduct, and then add to 96-well microplate for analysis.
11. Measure absorbance immediately on a microplate reader at OD=532 nm for colorimetric assay and RFU at Ex/Em = 532/553 nm for fluorometric assay (*see* **Note 9**).
12. Use the reader software or Excel to plot these values and curve fit. Calculate the amount of MDA in the organoids according to the standard curve.

3.3.4 Iron Assay

An iron assay protocol (based on typical organoids culture) is described in detail as follows [16]:

1. Prepare the necessary reagents in advance.
2. Prepare colorimetric assay standard curve dilution.
3. Harvest the number of organoids needed for each assay (initial recommendation = 10 ng), and repeat 3 times for each sample.
4. Wash organoids in cold PBS.
5. Homogenize organoids in 4–10 volumes of Iron Assay Buffer with a Dounce homogenizer sitting on ice, with 10–15 passes.
6. Centrifuge at 16,000 g for 10 min to remove insoluble materials.
7. Collect supernatant and transfer to a clean tube. Keep on ice.
8. Set up Reaction wells: Standard wells = 100 μL standard dilutions; Sample wells = 2–50 μL samples (adjust volume to 100 μL/well with Iron Assay Buffer).
9. Add 5 μL Iron Reducer to each standard well.
10. Add 5 μL of Assay Buffer to each sample.
11. Mix and incubate standards and samples at 37 °C for 30 min.
12. Add 100 μL Iron Probe to each well containing the iron standards and test samples.
13. Mix and incubate at 37 °C for 60 min protected from light.

14. Measure output immediately on a colorimetric microplate reader (OD=593 nm).
15. Use the reader software or Excel to plot these values and curve fit. Calculate the content of Fe^{2+} in the organoids according to the standard curve.

3.3.5 GSH Assay

A GSH assay protocol (based on typical organoids culture) is described in detail as follows [16]:

1. Prepare the necessary reagents in advance.
2. Prepare fluorometric assay standard curve dilution.
3. Harvest the number of organoids needed for each assay (initial recommendation = 100 mg), and repeat 3 times for each sample.
4. Wash organoids in cold PBS.
5. Rapidly homogenize tissue with 300 μL 5% SSA solution. Vortex vigorously and keep on ice for 10 min (*see* **Note 10**).
6. Centrifuge samples at 12,000 g at 4 °C for 20 min.
7. Collect supernatant and transfer to a clean tube. Keep on ice.
8. Dilute the samples 20–40-fold with GSH Assay Buffer.
9. Prepare Background control: add 100 μL GSH Assay Buffer and 200 μL 5% SSA to 100 μL distilled water. Vortex each sample and leave for 10 min on ice. Centrifuge at 12,000 g at 4 °C for 10 min. Collect the supernatant and keep on ice.
10. Set up Reaction wells: Standard wells =20 μL standard dilutions; Sample wells = 2–10 μL samples (adjust volume to 20 μL/well with GSH Assay Buffer); Sample background controls wells: add the same volume of diluted samples to designated wells (adjust volume to 20 μL/well with GSH Assay Buffer).
11. Prepare a 100-fold Dilution of Enzyme Mix A (i.e., Dilute 2 μL Enzyme Mix A stock solution with 198 μL GSH Assay Buffer). Mix well and keep on ice (*see* **Note 11**).
12. Prepare the Reaction Mix and Background Mix. Add 80 μL to the appropriate wells.
13. Measure fluorescence (Ex/Em = 535/587 nm) in kinetic mode at room temperature for 40–60 min.
14. Use the GSH standard curve to obtain the corresponding amounts of GSH in samples.

4 Notes

1. Biological Materials: Surgical resections or biopsies may be used for research purposes. Primary human tissue materials used in research must comply with all applicable institutional and governmental regulations. Informed consent must be obtained from all participants prior to the collection of tissue samples.
2. It is recommended to obtain tissue pieces from viable cancer tissue, which is typically pink in color, rather than necrotic tissue, which may appear yellowish and has poor organoid-forming efficiency. During transport to the laboratory, it is important to store tissue pieces in ice-cold basic medium containing 10 μM Y-27632 or, as an alternative, ice-cold PBS to minimize cell death and increase organoid outgrowth. Tissue pieces should be collected and stored in media as soon as possible, ideally in either the operating room or the pathological laboratory. These measures can improve cell viability.
3. It is crucial to closely monitor organoid cultures in the first few days after plating for potential bacterial and fungal infections. In case of such infections, 5 M NaOH (or another suitable disinfectant) should be immediately added to the infected wells, left for 1 h, and then the contents should be removed. Washing the contaminated well after removing the organoid material, followed by disinfection with 70% (vol/vol) ethanol and replacement of the culture plate lid, can minimize the chances of infection spreading to other well.
4. When using the CellTiter-Glo® 3D Reagent, it is important to handle the bottle seal with care to avoid introducing ATP contamination.
5. Some 3D cell culture methods can generate large amounts of biomass in a single well, which may impact assay performance. This typically occurs around 10 μM ATP, which is the upper limit of the assay linearity. If the biomass and/or ATP is in excess, we recommend either using smaller microtissue samples (if possible) or diluting the mixture of sample and reagent before recording luminescence.
6. Optimal assay performance is achieved when the CellTiter-Glo® 3D Reagent is thoroughly mixed with the cultured cells. Inefficient mixing can result in under-reporting of the amount of ATP present and may yield inaccurate results. We recommend vigorously shaking the mixture for 5 min and allowing a total of 30 min between reagent addition and luminescence recording. Other factors that can affect reagent mixing include the force of delivery, sample volume, and dimensions of the well. The method used to add CellTiter-Glo® 3D Reagent to

the assay plates can also impact the degree of reagent mixing required. Automated pipetting devices that use a greater or lesser force of fluid delivery may require different degrees of subsequent mixing. Complete reagent mixing in 96-well plates can be achieved using orbital plate shaking devices built into many luminometers and the recommended 5-min shaking time. We recommend considering these factors and optimizing assay conditions as necessary for your specific application.

7. The optimal detection instrument settings can vary depending on the manufacturer. As a guideline, we recommend using an integration time of 0.25–1 s per well. An uneven luminescent signal within plates can be caused by temperature gradients, uneven cell seeding, or edge effects in multiwell plates.
8. Occasionally, samples may exhibit turbidity, which can be eliminated by filtering them through a 0.2 μM filter. TBA can also react with other compounds in samples, resulting in the formation of other colored compounds. However, these should not interfere with the quantitation of the TBA-MDA adduct.
9. For fluorometric assays, we recommend setting the instrument sensitivity to high with a slit width of 5 nm.
10. GSH is sensitive to oxidation, so acidification of samples with SSA should be carried out as quickly as possible.
11. It is not recommended to store the Diluted Enzyme Mix A. Instead, prepare fresh dilutions as needed.

References

1. Dixon SJ, Lemberg KM, Lamprecht MR et al (2012) Ferroptosis: an iron-dependent form of nonapoptotic cell death. Cell 149:1060–1072
2. Guo J, Zhou Y, Liu D et al (2022) Mitochondria as multifaceted regulators of ferroptosis. Life Metab 1:134–148
3. Stockwell BR, Friedmann Angeli JP, Bayir H et al (2017) Ferroptosis: a regulated cell death nexus linking metabolism, redox biology, and disease. Cell 171:273–285
4. Tang D, Chen X, Kang R et al (2021) Ferroptosis: molecular mechanisms and health implications. Cell Res 31:107–125
5. Sato T, Vries RG, Snippert HJ et al (2009) Single Lgr5 stem cells build crypt-villus structures in vitro without a mesenchymal niche. Nature 459:262–265
6. Rossi G, Manfrin A, Lutolf MP (2018) Progress and potential in organoid research. Nat Rev Genet 19:671–687
7. Hu H, Gehart H, Artegiani B et al (2018) Long-term expansion of functional mouse and human hepatocytes as 3D organoids. Cell 175(1591–606):e19
8. Qian X, Nguyen HN, Song MM et al (2016) Brain-region-specific organoids using mini-bioreactors for modeling ZIKV exposure. Cell 165:1238–1254
9. Fujii M, Matano M, Toshimitsu K et al (2018) Human intestinal organoids maintain self-renewal capacity and cellular diversity in niche-inspired culture condition. Cell Stem Cell 23:787–93.e6
10. Morizane R, Lam AQ, Freedman BS et al (2015) Nephron organoids derived from human pluripotent stem cells model kidney development and injury. Nat Biotechnol 33: 1193–1200
11. Broutier L, Mastrogiovanni G, Verstegen MM et al (2017) Human primary liver cancer-derived organoid cultures for disease modeling and drug screening. Nat Med 23:1424–1435
12. Shinozawa T, Kimura M, Cai Y et al (2021) High-Fidelity drug-induced liver injury screen

using human pluripotent stem cell-derived organoids. Gastroenterology 160:831–46.e10

13. Chen X, Comish PB, Tang D et al (2021) Characteristics and biomarkers of Ferroptosis. Front Cell Dev Biol 9:637162
14. Chen F, Cai X, Kang R et al (2023) Autophagy-dependent ferroptosis in cancer. Antioxid Redox Signal. https://doi.org/10.1089/ars.2022.0202. Epub ahead of print
15. Chen X, Comish PB, Tang D, Kang R (2021) Characteristics and Biomarkers of Ferroptosis. Front Cell Dev Biol. 9:637162
16. Deng F, Zhao BC, Yang X et al (2021) The gut microbiota metabolite capsiate promotes Gpx4 expression by activating TRPV1 to inhibit intestinal ischemia reperfusion-induced ferroptosis. Gut Microbes 13:1–21
17. Mun SJ, Ryu JS, Lee MO et al (2019) Generation of expandable human pluripotent stem cell-derived hepatocyte-like liver organoids. J Hepatol 71:970–985
18. Olgasi C, Cucci A, Follenzi A (2020) iPSC-derived liver organoids: a journey from drug screening, to disease modeling, arriving to regenerative medicine. Int J Mol Sci 21:6215
19. Dutta D, Heo I, Clevers H (2017) Disease modeling in stem cell-derived 3D organoid systems. Trends Mol Med 23:393–410
20. Lancaster MA, Knoblich JA (2014) Organogenesis in a dish: modeling development and disease using organoid technologies. Science 345:1247125
21. Kleinman HK, Martin GR (2005) Matrigel: basement membrane matrix with biological activity. Semin Cancer Biol 15:378–386
22. Kratochvil MJ, Seymour AJ, Li TL et al (2019) Engineered materials for organoid systems. Nat Rev Mater 4:606–622
23. Aisenbrey EA, Murphy WL (2020) Synthetic alternatives to matrigel. Nat Rev Mater 5:539–551
24. Si-Tayeb K, Noto FK, Nagaoka M et al (2010) Highly efficient generation of human hepatocyte-like cells from induced pluripotent stem cells. Hepatology 51:297–305
25. Drost J, Clevers H (2018) Organoids in cancer research. Nat Rev Cancer 18:407–418
26. Meng Y, Sun J, Zhang G et al (2022) The three-dimension preclinical models for ferroptosis monitoring. Front Bioeng Biotechnol 10:1020971
27. Ben-David U, Ha G, Tseng YY et al (2017) Patient-derived xenografts undergo mouse-specific tumor evolution. Nat Genet 49:1567–1575
28. Li M, Izpisua Belmonte JC (2019) Organoids - preclinical models of human disease. N Engl J Med 380:569–579
29. Nuciforo S, Fofana I, Matter MS et al (2018) Organoid models of human liver cancers derived from tumor needle biopsies. Cell Rep 24:1363–1376
30. Sun L, Wang Y, Cen J et al (2019) Modelling liver cancer initiation with organoids derived from directly reprogrammed human hepatocytes. Nat Cell Biol 21:1015–1026
31. Tang D, Kroemer G (2020) Ferroptosis. Curr Biol 30:R1292-r7
32. Lorenzato A, Magrì A, Matafora V et al (2020) Vitamin C restricts the emergence of acquired resistance to EGFR-targeted therapies in colorectal cancer. Cancers 12:685
33. Guo J, Duan L, He X et al (2021) A combined model of human iPSC-derived liver organoids and hepatocytes reveals ferroptosis in DGUOK mutant mtDNA depletion syndrome. Adv Sci (Weinh) 8:2004680
34. Ouyang S, Li H, Lou L et al (2022) Inhibition of STAT3-ferroptosis negative regulatory axis suppresses tumor growth and alleviates chemoresistance in gastric cancer. Redox Biol 52:102317
35. Zhang Q, Deng T, Zhang H et al (2022) Adipocyte-derived exosomal MTTP suppresses Ferroptosis and promotes chemoresistance in colorectal cancer. Adv Sci (Weinh) 9:e2203357
36. Zou Y, Zheng S, Xie X et al (2022) N6-methyladenosine regulated FGFR4 attenuates ferroptotic cell death in recalcitrant HER2-positive breast cancer. Nat Commun 13:2672
37. Jiang X, Stockwell BR, Conrad M (2021) Ferroptosis: mechanisms, biology and role in disease. Nat Rev Mol Cell Biol 22:266–282
38. Stockwell BR (2022) Ferroptosis turns 10: emerging mechanisms, physiological functions, and therapeutic applications. Cell 185:2401–2421
39. Chen X, Li J, Kang R et al (2021) Ferroptosis: machinery and regulation. Autophagy 17:2054–2081
40. Zheng J, Conrad M (2020) The metabolic underpinnings of Ferroptosis. Cell Metab 32:920–937

Chapter 12

Analysis of Protein Degradation in Ferroptosis

Zhuojun Zhang and Lili Jiang

Abstract

The ubiquitin-proteasome system (UPS) is a highly conserved cellular mechanism that degrades and recycles proteins in eukaryotic cells. It involves the tagging of specific target proteins with ubiquitin, a small regulatory protein, which marks them for degradation by the proteasome, a large protein complex that acts as a molecular shredder. Dysfunction of the UPS has been implicated in a wide range of diseases, including cancer, neurodegenerative disorders, and viral infections. Therefore, targeting the UPS has become an attractive therapeutic strategy for many diseases. Ferroptosis is an iron-dependent cell death process that is regulated by multiple levels, including protein degradation. In this chapter, we introduce the detection of UPS-mediated protein degradation in ferroptosis using several techniques such as western blotting, co-immunoprecipitation, in vitro ubiquitination assay, and proteasome assay.

Key words Protein degradation, Ferroptosis, Ubiquitin-proteasome system

1 Introduction

The ubiquitin-proteasome system (UPS) is a crucial pathway responsible for the selective degradation of intracellular proteins. It involves the attachment of a small protein, ubiquitin, to a specific lysine residue on the target protein, marking it for degradation by the proteasome. This process is facilitated by a series of enzymatic reactions catalyzed by E1, E2, and E3 enzymes [1, 2]. Ubiquitination can occur on single or multiple lysine residues, leading to different types of ubiquitin linkages and various fates for the target protein [3]. Once marked with ubiquitin, the protein is recognized by the proteasome, which breaks it down into short peptides. The UPS process is reversible through deubiquitination, where ubiquitin is removed from the target protein by deubiquitinating enzymes (DUBs) [2].

Maintaining a balance between ubiquitination and deubiquitination is essential for proper cellular function, and an imbalance can lead to various diseases [4]. Recent studies have linked protein degradation to ferroptosis, a form of oxidative cell death

Guido Kroemer and Daolin Tang (eds.), *Ferroptosis: Methods and Protocols*, Methods in Molecular Biology, vol. 2712,
https://doi.org/10.1007/978-1-0716-3433-2_12,

characterized by iron and lipid peroxide accumulation [5–7]. The UPS plays a role in the degradation of proteins during ferroptosis, including solute carrier family 7 member 11 (SLC7A11), voltage-dependent anion channel (VDAC)-2/3, and lactotransferrin (LTF) [8]. The deubiquitinating enzyme OTU domain-containing ubiquitin aldehyde-binding protein 1 (OTUB1) stabilizes SLC7A11, leading to resistance to ferroptosis [9, 10], while increased ubiquitination and degradation of VDAC2/3 counteract the pro-ferroptosis effect of erastin through the upregulation of E3 ligase NEDD4 levels [11]. Additionally, the deubiquitinating enzyme USP35 maintains the stability of ferroportin against erastin- or RSL3-induced ferroptosis by reducing its ubiquitination level [12].

Examining protein degradation in the context of ferroptosis can yield valuable insights into cellular metabolism and regulatory mechanisms, as well as potential interventions for conditions involving dysregulated ferroptosis. This chapter provides an overview of methods for investigating the UPS in cultured mammalian cells during ferroptosis.

2 Materials

1. HT1080 or PANC1 cells.
2. Cell culture medium: Dulbecco's modified Eagle's medium (DMEM) supplemented with 10% fetal bovine serum (FBS).
3. Cell culture dish: Nunc™ EasYDish™ Dishes (ThermoFisher, #150466 for 100 mm, #150462 for 60 mm).
4. 1× Phosphate Buffered Saline (PBS): To preparel L of 10× PBS, add 80 g sodium chloride (NaCl), 2 g potassium chloride (KCl), 14.4 g sodium phosphate, dibasic (Na_2HPO_4) and 2.4 g potassium phosphate, monobasic (KH_2PO_4) to 1 L ddH_2O. Adjust pH to 7.2–7.4. Dilute 10× PBS by a ratio of 1: 10 with ddH_2O, mix well, and adjust the volume to prepare the 1× PBS.
5. Cycloheximide (CHX) solution: To prepare a 10 mM stock solution of CHX, add 5 mg of CHX to 1.777 mL of DMSO and mix well. To achieve a final working concentration of 200 μM, use a 1:50 dilution ratio by adding 200 μL of the stock solution to 10 mL of cell-culture medium and mix well.
6. MG132 (10 mM, dissolved in DMSO) solution: Dilute the MG132 stock solution with cell-culture medium according to the specific experimental requirements. The recommended concentrations are 10 μM for 6 hr, 20 μM for 2 hr, and 1–5 μM for 24 hr. However, the optimal concentration may

vary depending on the desired duration of action and experimental conditions.

7. Pierce BCA Protein Assay Kit (ThermoFisher, # 23227).
8. Lysis buffer: To prepare 10 mL of lysis buffer, add 1 mL of 10× RIPA lysis buffer (Cell Signaling, #9806), 100 μL of protease inhibitors (KeyGEN, #KGP602), 100 μL of phosphatase inhibitors (KeyGEN, #KGP603), and 100 μL of PMSF (FudeBio, #FD0100) to ddH2O to a final volume of 10 mL and mix well. This buffer should be freshly prepared.
9. 6× loading buffer (Beyotime, #P0015F).
10. SDS-polyacrylamide gel electrophoresis (SDS-PAGE): To prepare the upper buffer, dissolve 12.12 g of Tris base and 0.8 g of SDS in ddH$_2$O. Adjust the pH to 6.8, then bring the final volume to 500 mL. To prepare the lower buffer, dissolve 90.83 g of Tris base and 2 g of SDS in ddH$_2$O. Adjust the pH to 8.8, then bring the final volume to 500 mL. To prepare a 30% acrylamide-N, N′-Methylenebisacrylamide (Acr-Bis) solution, dissolve 262.8 g of acrylamide and 7.2 g of bis-acrylamide in ddH$_2$O to a final volume of 900 mL. To prepare a 10% ammonium persulfate solution (APS), dissolve 0.1 g of APS in 1 mL of ddH$_2$O. Prepare only as much 10% APS as needed for immediate use.

Two types of gels are used in SDS-PAGE: the stacking gel and the separation gel. To prepare a stacking gel, mix 0.8 mL of 30% Acr-Bis, 1.5 mL of lower buffer, 60 μL of 10% APS, and 6 μL of tetramethylethylenediamine (TEMED) in a final volume of 6 mL with ddH$_2$O. To prepare a 9% separation gel, mix 6 mL of 30% Acr-Bis, 3 mL of lower buffer, 106.7 μL of 10% APS, and 10.1 μL of TEMED in a final volume of 20 mL with ddH$_2$O. Divide this volume into two pieces of SDS-PAGE gels for a vertical electrophoresis system (Bio-Rad, #1658004). The concentration of the separation gel should be adjusted according to the molecular weight of the target protein by varying the amount of 30% Acr-Bis used (*see* **Notes 1** and **2**).

11. Running buffer: To prepare 4 L 10× running buffer, add 576 g of glycine, 120 g of Tris base, and 40 g of SDS to 4 L of ddH$_2$O. Dilute 10× running buffer by a ratio of 1:10 with ddH$_2$O, mix well and adjust the volume to prepare the 1× running buffer.
12. Transfer buffer: To prepare the 10× transfer buffer, add 580 g of glycine and 116 g of Tris base to 4 L of ddH$_2$O. Dilute the 10× transfer buffer by a ratio of 1:10 with ddH$_2$O, mix well, and adjust the volume to prepare the 1 × transfer buffer.
13. TBST: To prepare 1 L TBST, add 1 mL of Tween-20 (0.1%) to 1 × Tris-buffered saline (TBS), mix well and adjust the final

volume to 1 L. To make 1 L of 10× TBS, dissolve 24 g Tris base and 80 g NaCl in 900 mL of ddH_2O and then adjust the pH to 7.4 and final volume to 1 L, dilute the 10× TBS by a ratio of 1:10 with ddH_2O, mix well and adjust the volume to prepare the 1× TBS.

14. Blocking buffer: To prepare 100 mL of blocking buffer, add 5 g of nonfat powdered milk to 100 mL of TBST.
15. Primary antibody: anti-SLC7A11 (Cell Signaling, #12691), anti-GPX4 (Abcam, #ab252833), or anti-K48-linkage Specific Polyubiquitin (Cell Signaling, #8081).
16. Primary buffer: To prepare 20 mL primary buffer, add 1.0 g bovine serum albumin (BSA) to 20 mL TBST and mix well.
17. HRP-linked secondary antibody: anti-rabbit IgG, HRP-linked Antibody (Cell Signaling, #7074) or anti-mouse IgG, HRP-linked Antibody (Cell Signaling, #7076).
18. The enhanced chemiluminescence (ECL): Western Blotting Luminol Reagent (Santa Cruz, #sc-2048), mainly used to detect high levels of protein.
19. 1× IP cell lysis buffer: add 1 mL 10× cell lysis buffer (Cell Signaling, #9803) to 9 mL ddH_2O, mix well. Add 100 μL of PMSF (FudeBio, #FD0100) before use.
20. Magnetic Beads: Protein A (Cell Signaling, #73778) for rabbit IgG pull down, and Protein G (Cell Signaling, #70024) for mouse IgG pull down.
21. 3× SDS sample buffer: Use the Blue Loading Buffer Pack (Cell Signaling, #7722) to prepare 3× SDS sample buffer, i.e., add 1/10 volume 30× DTT reducing regent to 1 volume of 3× SDS loading buffer provided in the pack.
22. Ultrasonic homogenizers (SONICS, #VCX130).
23. Vortex mixer (SciLogex, #SCI-VS).
24. Dry Bath Incubator (MIULAB, #DKT-100).
25. Horizontal shaker (Kylin-Bell, #TS-2000A).
26. Vertical electrophoresis system: Mini-PROTEAN® Tetra Vertical Electrophoresis Cell for Mini Precast Gels (Bio-Rad, #1658004), supplemented with PowerPac™ Basic Power Supply (Bio-Rad, #1645050).
27. Magnetic separator: 6-Tube Magnetic Separation Rack (Cell Signaling, #7017).
28. Film (Kodak, #XBT) or digital imaging system: Alliance Q9 Advanced Chemiluminescence Imager (UVITECs, Cleaver Scientific).
29. The reaction of in vitro ubiquitination assay: Ubiquitin (R&D, #U-100H-10 M), MgATP solution (R&D, #B-20), E1

enzyme (R&D, #E-304-050), E2 enzyme (R&D, #E2–616-100), E3 ligase (R&D, #E3–411-025). To prepare 10× E3 Ligase Reaction Buffer, mix 500 mM 4-(2-Hydroxyethyl) piperazine-1-ethanesulfonic acid (HEPES), pH 8.0, 500 mM NaCl, and 10 mM Tris (2-carboxyethyl) phosphine (TCEP).

30. Water bath (Grant Instruments™ JBA18, Fisher Scientific).
31. Proteasome 26S Degradation Activity Kit, e.g., Abcam #ab139466, #ab107921 or Enzo Life Sciences #BML-PW8950–0001 can be recommended.
32. Proteasome 20S Degradation Activity Kit, e.g., Sigma-Aldrich #MAK172, Abcam #ab112154 or AAT Bioquest #13456 can be recommended.
33. Scintillation counter (MicroBeta2, PerkinElmer) or fluorescence microplate reader (GM3000, Promega).

3 Methods

3.1 Monitoring UPS Protein Degradation Using Western Blotting

Western blotting (WB) is a widely used laboratory technique to detect and quantify specific proteins in a complex mixture of proteins, with a broad range of applications in molecular biology, including disease diagnosis, protein-protein interaction studies, and post-translational modification detection [13, 14]. The technique involves separating the proteins by size through gel electrophoresis, followed by transfer onto a nitrocellulose (NC) or polyvinylidene fluoride (PVDF) membrane. The membrane is then probed with specific antibodies against the target protein of interest, which are often labeled with a secondary antibody conjugated with an enzyme or fluorescent dye for visualization.

To investigate protein degradation via the UPS pathway in the context of ferroptosis, it is recommended to use WB to analyze the ubiquitination levels of proteins within a sample. Cells can be treated with cycloheximide (CHX), a translation inhibitor that prevents the elongation step of protein synthesis by binding to the ribosome, to block protein synthesis and enable the detection of changes in protein levels over time. This approach is useful for studying the turnover of specific proteins targeted for degradation by the UPS pathway and for determining the kinetics of degradation in response to various treatments or conditions [15, 16]. Furthermore, the use of a proteasome inhibitor such as MG132 can block the proteasomal degradation of ubiquitinated proteins, leading to the accumulation of ubiquitinated proteins within the cells [17, 18]. Changes in the levels of ubiquitinated proteins can indicate changes in the activity of the UPS pathway, as alterations in the pathway's components or activity can affect the levels of ubiquitinated proteins.

1. Place erastin or RSL3-treated HT1080 or PANC1 cell culture dish on ice and remove the medium from the dish. Wash the cells with cold 1× PBS (pH 7.2–7.4).
2. Add cell lysis buffer containing protease inhibitors, phosphatase inhibitors, and PMSF to the dish. Scrape the cells and transfer the extract to a 1.5 mL microcentrifuge tube.
3. Keep the tube on ice and sonicate the extract with an ultrasonic homogenizer for 10–15 sec.
4. Centrifuge the extract at 19,000 G-force for 10 min at 4 °C.
5. Transfer the supernatant to a new microcentrifuge tube and store it on ice.
6. Use the BCA Protein Assay kit to quantify protein concentration. Adjust the concentration using the lysis buffer.
7. Add 6× loading buffer to the extract and boil the mixture at 95 °C for 5 min.
8. Load equal amounts of protein and molecular weight markers into an SDS-PAGE gel.
9. Run the gel in a running buffer with a vertical electrophoresis system, at 80 V for 5 min, then increase the voltage to 120 V and complete the run in about 40–60 min (*see* **Note 3**).
10. Transfer the proteins from the SDS-PAGE gel to a PVDF membrane in a transfer buffer, at a constant current of 285 mA for 150 min (*see* **Notes 4–6**).
11. Incubate the membrane with a blocking buffer for 1 hr on a horizontal shaker at room temperature.
12. Wash the membrane three times for 5 min with TBST.
13. Incubate the membrane with the primary antibody (diluted in a primary buffer according to the product instruction, e.g., anti-SLC7A11 antibody 1:1000, anti-K48 ubiquitin 1:1000) overnight at 4 °C.
14. Wash the membrane three times for 5 min with TBST.
15. Incubate the membrane with the species-appropriate HRP-linked secondary antibody (diluted with 5% nonfat powdered milk according to the product instruction) for 1 hr at room temperature.
16. Wash the membrane three times for 5 min with TBST.
17. Prepare the ECL according to the product instruction and mix it with the membrane for 1 min (*see* **Note 7**).
18. Use a film or digital imaging system to capture the chemiluminescent signal and use image analysis software to quantify the protein levels (*see* **Note 8**).

3.2 Monitoring UPS Protein Degradation Using Co-immunoprecipitation

Co-immunoprecipitation (Co-IP) is a laboratory technique used to identify protein-protein interactions. The technique involves isolating a target protein from a complex mixture of proteins, such as a cell lysate, using an antibody that is bound to a solid support, such as protein A or protein G agarose beads. Once the antibody-protein complex is formed, the sample is washed to remove any non-specifically bound proteins. Then, a second antibody specific to the interacting protein is added, which binds to the first antibody and pulls down any interacting proteins that are bound to the target protein. The resulting protein-protein interaction can be detected by WB, which allows for the visualization of the interacting protein [19].

Co-IP can be used to investigate protein degradation by the UPS pathway. By examining the interaction between a target protein and specific proteins involved in the degradation pathway, such as E3 ubiquitin ligases or ubiquitin itself, it is possible to gain insights into how the target protein is regulated and degraded. By comparing the levels of the protein of interest and its associated proteins before and after treatment with UPS or DUB inhibitors, researchers can determine if the protein is being degraded by the UPS pathway [20, 21].

1. Place the cell culture dish on ice. Remove the medium and wash the cells with 1× PBS (pH 7.2–7.4).
2. Add 0.5 mL of ice-cold 1× IP cell lysis buffer to the cell culture dish (100 mm) and incubate on ice for 5 min.
3. Scrape the cells out of the dish and transfer the extract to a 1.5 mL microcentrifuge tube.
4. Sonicate the sample three times for 5 sec each time. Centrifuge at 19,000 G-force for 10 min at 4 °C. Transfer the supernatant to a new microcentrifuge tube and keep it on ice.
5. Pre-clear the lysate to reduce non-specific protein binding to the magnetic beads. Briefly vortex the stock tube of magnetic beads to resuspend them. Absorb 20 μL of bead slurry into a clean microcentrifuge tube. Put the tube on a magnetic separator for 10–15 sec. When the solution is clear, carefully remove the buffer. Vortex briefly after adding 500 μL of 1× IP cell lysis buffer to the magnetic bead precipitate to wash the beads. Put the tube on the magnetic separator. When the solution is clear, carefully remove the buffer. Repeat the washing step once more.
6. Absorb 200 μL of cell lysate into 20 μL of pre-washed magnetic beads. Incubate in rotation for 20 min at room temperature.
7. Use a magnetic separator to separate the beads from the lysate, transfer the pre-cleared lysate to a new tube, then remove the magnetic bead precipitate.

8. Dilute the primary antibody in a primary buffer according to the ratio recommended in the product instruction (e.g., add 120 ng of anti-K48 ubiquitin to the cell lysate containing 100 μg of protein). Absorb the diluted primary antibody into 200 μL of cell lysate. Incubate in rotation overnight at 4 °C to form an immunocomplex.
9. Pre-wash magnetic beads as described in **step 5**.
10. Transfer the immunocomplex solution to the tube including the pre-washed magnetic bead pellet. Incubate in rotation for 20 min at room temperature.
11. Use the magnetic separator to precipitate beads. Wash beads with 500 μL of 1× IP cell lysis buffer five times, keeping them on ice.
12. Resuspend the precipitate with 20–40 μL of 3× SDS sample buffer. Vortex briefly and microcentrifuge to precipitate.
13. Boil the sample at 95–100 °C for 5 min.
14. Use a magnetic separator to precipitate beads. Absorb the supernatant (i.e., the sample) into a new tube.
15. Load the sample onto the SDS-PAGE gel and analyze the sample by WB (*see* **Notes 9–11**).

3.3 Monitoring UPS Protein Degradation Using In Vitro Ubiquitination Assay

The in vitro ubiquitination assay is a laboratory technique used to detect the direct ubiquitination of target proteins and replicate the process of the UPS. The technique involves incubating purified E1, E2, and E3 enzymes, along with ubiquitin and a target protein or substrate, in the presence of ATP. This facilitates the formation of an isopeptide bond between the ubiquitin molecule and a lysine residue on the target protein, resulting in the attachment of the ubiquitin to the target protein.

The in vitro ubiquitination assay is useful for confirming the production of ubiquitination products, validating the ubiquitinated substrate, or identifying the autoubiquitination of E3 ligases. Methods such as WB or mass spectrometry can be employed to analyze the reaction products [22].

1. Prepare a 25 μL reaction by combining the following components in a microcentrifuge tube (*see* **Note 12**):
   (a) 2.5 μL of 10× E3 Ligase Reaction Buffer.
   (b) 1 μL of ubiquitin to a final concentration of 100 μM.
   (c) 2.5 μL of MgATP solution to a final concentration of 10 mM.
   (d) 0.5 μL of E1 enzyme to a final concentration of 100 nM.
   (e) 1 μL of E2 enzyme to a final concentration of 1 μM.

(f) 1 μM of E3 ligase and 5–10 μM of a target protein or substrate.

(g) Add ddH_2O to adjust the total volume to 25 μL.

2. Incubate the reaction at 37 °C in a water bath for 30–60 min.
3. Terminate the reaction by adding 25 μL of 1× SDS-PAGE sample buffer.
4. To analyze ubiquitin conjugation reaction products, follow these steps:

(a) Separate the reaction products by SDS-PAGE and visualize the resulting gel with Coomassie blue staining to identify the presence of ubiquitinated products.

(b) Conduct a WB using anti-ubiquitin and/or anti-substrate antibodies to confirm the presence of ubiquitinated substrate, or anti-E3 ligase antibodies to identify E3 ligase autoubiquitination.

3.4 Monitoring UPS Protein Degradation Using Proteasome Assay

The proteasome is a multi-subunit complex responsible for the degradation of most intracellular proteins targeted by the UPS [23]. The 26S proteasome undergoes several enzymatic and non-enzymatic steps to rapidly degrade ubiquitinated proteins. These steps include binding ubiquitinated substrates to the 19S particle, opening the gated substrate entry channel into the 20S particle, disassembling the Ub chain, ATP hydrolysis, substrate unfolding and translocation, and proteolysis within the 20S particle. Accurately measuring each of these steps is crucial to fully understand the physiological regulation of proteasome function and the impact of diseases and drugs.

Depending on the labeling method used for the substrate, different techniques can be employed to measure proteasome activity. These include using fluorogenic peptides, radiolabeled peptides, or colorimetric substrates. The degree of proteasome activity can be determined by measuring the release of fluorescent, radioactive, or colored products using a fluorometer, scintillation counter, or spectrophotometer, respectively.

3.4.1 Proteasome 26S Assay

The Proteasome 26S assay is a commonly used method to measure the activity of the 26S proteasome in protein degradation via the UPS [24]. This assay typically employs a Proteasome 26S Degradation Activity Kit, which includes purified 26S proteasome, fluorescently labeled ubiquitinated substrates, a degradation buffer, and a control inhibitor to ensure assay specificity. To conduct the assay, the substrate is incubated with the proteasome in the degradation buffer. Active proteasomes recognize and degrade the substrate, resulting in the release of the fluorescent label. The fluorescence can then be quantified using a fluorometer to determine the proteasome's activity.

In addition to fluorescent labeling, other methods such as colorimetric and radiometric assays also exist for measuring proteasome 26S activity [25, 26]. These assays employ different labeling methods for the substrates. Measuring proteasome activity accurately is crucial for understanding the physiological regulation of proteasome function and the impact of diseases and drugs on this system.

1. In a microcentrifuge tube, combine radio-labeled ubiquitinated protein conjugate at a concentration of 50-100 nM/ 5000-15,000 cpm, MgATP, and 26S proteasome.
2. Incubate the reaction mixture at 37 °C for 0–90 min.
3. At the desired time points, remove 10 μL of the reaction mixture and add it to a quenching buffer containing 600 μL of 10% trichloroacetic acid (TCA) and 200 μL of 5% BSA as a carrier.
4. Vortex the mixture and then place it on ice for 15 min.
5. Centrifuge the mixture at 4 °C for 10 min in a microcentrifuge at 20,000 G-force.
6. Count 650 μL of the supernatant in a scintillation counter.
7. To analyze the data, plot the increase in TCA-soluble counts versus time (*see* **Note 13**).

3.4.2 Proteasome 20S Assay

Proteasome 20S assays are commonly used to assess the activity of the 20S proteasome in protein degradation, especially non-ubiquitin-mediated proteolysis [27]. The Proteasome 20S Activity Kit is primarily employed to measure the activity of the 20S proteasome or screen inhibitors in solution or cultured cells.

One common method involves using LLVY-R110 as a fluorogenic substrate. When the proteasome cleaves this substrate, it produces strongly green fluorescent R110 that can be monitored fluorometrically at 520–530 nm, with excitation at 480–500 nm. Enhanced fluorescence intensity indicates greater proteasome 20S activity. Some kits offer purified 20S proteasome and are designed to measure the activity of specific proteasome subunits, such as the Chymotrypsin-like activity of the ß5/PSMB5 subunit using Suc-LLVY-AMC, Caspase-like activity of the ß1/PSMB6 subunit using LLE-AMC, and Chymotrypsin-like activity of other subunits using WLA-AMC. In these assays, AMC is used as a fluorescent reporter molecule to indicate the activity of the proteasome enzyme [28, 29].

1. Plate cells in a growth medium at a density of 80,000 cells per well in a 96-well plate (scale as needed in all indicated steps). Incubate the plate overnight to allow cells to adhere and grow.
2. Allow all reagents to come to room temperature and briefly centrifuge vials before opening.

3. Treat cells with 10 μL of 10× test compound in PBS or the desired buffer. Ensure that the test compound is well mixed before use.
4. Prepare a blank well with medium without cells and add the corresponding amount of buffer used to treat the cells.
5. Incubate the cell plates in a 5% CO_2 incubator at 37 °C for the desired duration.
6. Prepare the Proteasome Assay Loading Solution by mixing 25 μL of 400× Proteasome LLVY-R110 Substrate Stock Solution with 10 mL of Assay Buffer. To prepare the Substrate Stock Solution, mix 25 μL of DMSO with a vial of Proteasome LLVY-R110 Substrate at a concentration of 400×.
7. Add 100 μL of the prepared loading solution to each well.
8. Incubate the plate for at least 1 hr at 37 °C or room temperature, protected from light. The optimal incubation time must be determined, and longer incubations may be required.
9. Measure the fluorescence intensity using a fluorescence plate reader with an excitation wavelength of 490 nm and an emission wavelength of 525 nm.
10. Record the results and analyze the data as needed.

4 Notes

1. Composition ratios of separation gels with different acrylamide concentrations (Table 1). Divide this volume into 2 pieces of SDS-PAGE gels for a vertical electrophoresis system (Bio-Rad, #1658004).
2. If you're researching a small molecular weight protein (<20 kDa), consider the following suggestions. To ensure that the target protein can be separated, prepare a 12–15% SDS-PAGE separation gel. For target proteins with a molecular weight between 1 and 10 kDa, using Tricine-SDS-PAGE is a

Table 1
Composition ratios of separation gels

	6%	7.5%	9%	10.5%	12%	13.5%	15%
ddH_2O (mL)	10.8	9.88	8.8	8	6.94	5.8	4.91
30% Acr-Bis (mL)	4	5	6	7	8	9	10
Lower buffer (mL)	5	5	5	5	5	5	5
10% APS (μL)	106.7	106.7	106.7	106.7	106.7	106.7	106.7
TEMED (μL)	10.1	10.1	10.1	10.1	10.1	10.1	10.1

better choice. Increase the ratio of the stacking gel to fully compress the protein sample before separation. This will help ensure optimal separation and accurate analysis of the protein sample.

3. Maintain optimal cooling during electrophoresis to prevent high temperatures from accelerating molecular movement and causing band dispersion.
4. Use 20% methanol in the transfer buffer, but avoid using SDS, which can interfere with protein binding to the membrane and affect the transfer efficiency.
5. Use a 0.22 μm PVDF membrane instead of a 0.45 μm membrane for improved protein binding and signal-to-noise ratio.
6. Start with a current of 250 mA for 15–30 min for effective transfer, but optimize these parameters based on the target protein's molecular weight, charge, and other characteristics to achieve the best results.
7. If the target protein has low levels, it is recommended to use a sensitive ECL reagent, e.g., Meilunbio® fg super sensitive ECL luminescence reagent (Meilunbio, #MA0186) can be used.
8. If you obtain low or no signal from your WB results, we recommend including a known positive control to confirm experimental results. If the amount of protein loaded is too low, it can be difficult to detect your protein of interest, particularly if it is less abundant in your sample. This may necessitate increasing the exposure time to detect the target protein on your blot. Alternatively, you can increase the loading protein or antibody concentrations. If you observe multiple bands or non-specific binding, we suggest using the dilution recommended on the product webpage and datasheet as a starting point for optimization. Proteases and phosphatases present in your sample can quickly degrade your protein. Adding protease and phosphatase inhibitors to your lysate can help slow or prevent protein degradation. The buffer used for the dilution of the primary antibody can also affect the background on your membrane. For example, a buffer such as non-fat dry milk can reduce the appearance of non-specific bands and overall background on the blot compared to BSA. However, milk may be too stringent for some antibodies and lead to a reduced target signal. Therefore, it is crucial to refer to the recommended WB protocol on the company's product webpage.
9. If you encounter low binding of the selecting antibody during IP, it is essential to verify the binding/specificity of the antibody to the antigen. You can do this by performing an ELISA or checking the binding of antibodies to the beads. If the antibodies are not captured and bound to the beads, the IP experiment will not work. Check the amount of beads and

sample volume. With reference to the capacity of different beads proposed in the package inserts, increasing the amount of beads or the concentration of your antibody during coupling can be helpful. If this doesn't work, you can also try increasing the incubation time or using another antibody.

10. Use input control (whole lysate) to verify WB is working and detect unsuccessful IP enrichment. Employ isotype control (matching the IgG subclass of the primary antibody) to avoid false positives. Choose protein A beads for rabbit antibodies and protein G beads for mouse antibodies to capture antibodies due to their different affinities for each IgG subclass.
11. When performing IP followed by WB, the denatured IgG light and heavy chains of the primary antibody can run at 25 and 50 kDa, potentially masking target proteins with similar molecular weights. To avoid this, either use antibodies from different host animals for the IP and WB or utilize a conformation-specific secondary antibody, such as mouse anti-rabbit IgG (Cell Signaling, #5127), if using antibodies from the same host animals.
12. Each E2 enzyme has specificity for a subset of E3 ligases, and the interactions between E2s and E3s are highly regulated. Researchers can either provide their own E3 ligase or purchase commercially available products from certain companies.
13. The TCA-soluble counts correspond to the amount of degraded substrate, and an increase in counts represents the proteasome's activity. The quenching buffer with TCA stops the proteasome activity, and the carrier protein (bovine serum albumin) helps to precipitate the protein, facilitating the removal of unincorporated labels and improving counting accuracy.

References

1. Tai HC, Schuman EM (2008) Ubiquitin, the proteasome and protein degradation in neuronal function and dysfunction. Nat Rev Neurosci 9:826–838
2. Neutzner M, Neutzner A (2012) Enzymes of ubiquitination and deubiquitination. Essays Biochem 52:37–50
3. Peng J, Schwartz D, Elias JE et al (2003) A proteomics approach to understanding protein ubiquitination. Nat Biotechnol 21:921–926
4. Sun T, Liu Z, Yang Q (2020) The role of ubiquitination and deubiquitination in cancer metabolism. Mol Cancer 19:146
5. Tang D, Kroemer G (2020) Ferroptosis. Curr Biol 30:R1292–R12R7
6. Dixon SJ, Lemberg KM, Lamprecht MR et al (2012) Ferroptosis: an iron-dependent form of nonapoptotic cell death. Cell 149:1060–1072
7. Xie Y, Hou W, Song X et al (2016) Ferroptosis: process and function. Cell Death Differ 23: 369–379
8. Wang Y, Liu Y, Liu J et al (2020) NEDD4L-mediated LTF protein degradation limits ferroptosis. Biochem Biophys Res Commun 531: 581–587
9. Liu T, Jiang L, Tavana O et al (2019) The Deubiquitylase OTUB1 mediates Ferroptosis via stabilization of SLC7A11. Cancer Res 79: 1913–1924

10. Chen X, Yu C, Kang R et al (2021) Cellular degradation systems in ferroptosis. Cell Death Differ 28:1135–1148
11. Yang Y, Luo M, Zhang K et al (2020) Nedd4 ubiquitylates VDAC2/3 to suppress erastin-induced ferroptosis in melanoma. Nat Commun 11:433
12. Tang Z, Jiang W, Mao M et al (2021) Deubiquitinase USP35 modulates ferroptosis in lung cancer via targeting ferroportin. Clin Transl Med 11:e390
13. Towbin H, Staehelin T, Gordon J (1979) Electrophoretic transfer of proteins from polyacrylamide gels to nitrocellulose sheets: procedure and some applications. Proc Natl Acad Sci U S A 76:4350–4354
14. Kurien BT, Aggarwal R, Scofield RH (2019) Protein extraction from gels: a brief review. Methods Mol Biol 1855:479–482
15. Qian H, Zhang N, Wu B et al (2020) The E3 ubiquitin ligase Smurf2 regulates PARP1 stability to alleviate oxidative stress-induced injury in human umbilical vein endothelial cells. J Cell Mol Med 24:4600–4611
16. Cui SZ, Lei ZY, Guan TP et al (2020) Targeting USP1-dependent KDM4A protein stability as a potential prostate cancer therapy. Cancer Sci 111:1567–1581
17. Xue W, Li X, Li W et al (2022) Intracellular CYTL1, a novel tumor suppressor, stabilizes NDUFV1 to inhibit metabolic reprogramming in breast cancer. Signal Transduct Target Ther 7:35
18. Hao X, Shiromoto Y, Sakurai M et al (2022) ADAR1 downregulation by autophagy drives senescence independently of RNA editing by enhancing p16(INK4a) levels. Nat Cell Biol 24:1202–1210
19. Bonifacino JS, Dell'Angelica EC, Springer TA (2001) Immunoprecipitation. Curr Protoc Mol biol chapter 10: unit 10.6
20. Yang Q, Yan D, Zou C et al (2022) The deubiquitinating enzyme STAMBP is a newly discovered driver of triple-negative breast cancer progression that maintains RAI14 protein stability. Exp Mol Med 54:2047–2059
21. de Sousa Leal AM, de Azevedo Medeiros LB, Munoz-Cadavid CO et al (2020) XPA deficiency affects the ubiquitin-proteasome system function. DNA Repair (Amst) 94:102937
22. Zhao Q, Liu L, Xie Q (2012) In vitro protein ubiquitination assay. Methods Mol Biol 876:163–172
23. Nandi D, Tahiliani P, Kumar A et al (2006) The ubiquitin-proteasome system. J Biosci 31:137–155
24. Jonsson E, Htet ZM, Bard JAM et al (2022) Ubiquitin modulates 26S proteasome conformational dynamics and promotes substrate degradation. Sci Adv 8:eadd9520
25. Bhattacharyya S, Renn JP, Yu H et al (2016) An assay for 26S proteasome activity based on fluorescence anisotropy measurements of dye-labeled protein substrates. Anal Biochem 509:50–59
26. Kisselev AF, Goldberg AL (2005) Monitoring activity and inhibition of 26S proteasomes with fluorogenic peptide substrates. Methods Enzymol 398:364–378
27. Ben-Nissan G, Sharon M (2014) Regulating the 20S proteasome ubiquitin-independent degradation pathway. Biomol Ther 4:862–884
28. Smith DM, Chang SC, Park S et al (2007) Docking of the proteasomal ATPases' carboxyl termini in the 20S proteasome's alpha ring opens the gate for substrate entry. Mol Cell 27:731–744
29. Kisselev AF, Kaganovich D, Goldberg AL (2002) Binding of hydrophobic peptides to several non-catalytic sites promotes peptide hydrolysis by all active sites of 20 S proteasomes. Evidence for peptide-induced channel opening in the alpha-rings. J Biol Chem 277:22260–22270

Chapter 13

Lipidomics Analysis in Ferroptosis

Zhi Lin and Minghua Yang

Abstract

Ferroptosis is a form of regulated cell death that occurs due to abnormal lipid metabolism. Lipids, which have been identified in over 45,000 different molecular species, play essential roles in modulating basic life processes. The process of ferroptosis is highly reliant on various lipid species, with polyunsaturated fatty acids (PUFAs) playing a central role in driving this process. Recent advances in mass spectrometry-based lipidomics have led to a surge in studies on ferroptosis. To explore the mechanism of lipid homeostasis in ferroptosis, the development of lipidomics techniques is critical. Currently, liquid chromatography-tandem mass spectrometry (LC-MS/MS) and gas chromatography-mass spectrometry (GC-MS) are the most widely used analytical techniques in lipidomics. These techniques offer deeper insights into the complex lipid mechanisms that underlie ferroptosis.

Key words Lipidomics, Ferroptosis, Mass spectrometry

1 Introduction

Ferroptosis is a complex form of iron-dependent regulated cell death driven by abnormal lipid metabolism [1, 2]. This process results in lipid peroxidation and subsequent plasma membrane rupture. Lipids are crucial components of organisms and play critical roles in modulating basic life processes, including cell membrane formation, energy storage, signaling, and maintenance of the epithelial barrier [3–6]. With a structural diversity estimated to be over 45,000, there are thousands of different molecular species of lipids in organisms, as documented in LIPID MAPS [7]. These lipids have various combinations of molecular backbones, polar head groups, acyl chains, double-bond positions, sn-positions, and stereo-centers. The process of ferroptosis is highly dependent on different types of lipid species that can originate from various sources. For example, polyunsaturated fatty acids (PUFAs) play a central role in driving ferroptosis [8–11], while exogenous monounsaturated fatty acids (MUFAs) inhibit ferroptosis [12, 13].

Guido Kroemer and Daolin Tang (eds.), *Ferroptosis: Methods and Protocols*, Methods in Molecular Biology, vol. 2712,
https://doi.org/10.1007/978-1-0716-3433-2_13,

Recent advances in mass spectrometry-based lipidomics have resulted in a rapid proliferation of studies on ferroptosis. Combining mass spectrometry imaging with high-resolution mass analysis offers a more in-depth understanding of the lipid mechanisms of ferroptosis. Mass spectrometry-based metabolomics approaches can detect and quantify thousands of metabolite features simultaneously [14–16]. High mass resolution enables confident assignment of lipid classes, such as phosphatidylcholine (PC), phosphatidylethanolamine (PE), phosphatidylserine (PS), phosphatidylglycerol (PG), phosphatidylinositol (PI), and phosphatidic acid (PA), as well as lipid species that have distinct combinations of fatty acyl (FA) moieties. Identifying individual FA chains is crucial for a better understanding of lipid biochemistry, which is increasingly recognized as playing significant roles in ferroptosis.

Mass spectrometry-based lipidomics detection methods can be classified into five categories [17]: Direct Infusion Mass Spectrometry (DIMS) for analyzing single lipid molecules, Liquid Chromatography-Mass Spectrometry (LC-MS) and Ion Mobility Mass Spectrometry (IM-MS) for analyzing complex lipid mixtures, Gas Chromatography-Mass Spectrometry (GC-MS) for analyzing volatile lipids, and Tandem Mass Spectrometry (MS/MS) for identifying lipid structures and performing quantitative analysis. Among these, LC-MS/MS is currently the most widely used analytical technique in lipidomics [18, 19], while GC-MS is commonly used for analyzing fatty acids and lipid metabolites, such as cholesterol, steroid hormones, and bile acids, which are important in disease development [20].

In conclusion, developing lipidomics techniques is crucial for exploring the mechanism of lipid homeostasis in ferroptosis. Mass spectrometry-based untargeted lipidomics has revealed the lipidome atlas during ferroptosis at the molecular species level. This chapter provides an overview of LC-MS/MS and GC-MS-based lipidomics methods, which are essential tools for understanding the complex lipid mechanisms of ferroptosis.

2 Materials

2.1 For LC-MS/MS

1. Reagents: MS-grade methanol; MS-grade acetonitrile; Grade 2-propanol; Formic Acid; Ammonium formate; Tert-butyl Methyl Ether (MTBE); SPLASH® LIPIDOMIX Internal Standards; HPLC Buffer A:acetonitrile water solution (acetonitrile: water = 6:4, v/v); HPLC Buffer B:acetonitrile isopropanol solution (acetonitrile: isopropanol = 1:9, v/v).
2. Liquid chromatography system: Ultra-high-performance liquid chromatography (UHPLC) system is used to separate the lipid molecules based on their physicochemical properties.

3. Mass spectrometer: The samples were separated by UHPLC and analyzed by a Q exactive mass spectrometer.
4. Electrospray ionization (ESI) source: This is used to ionize the separated lipid molecules in the mass spectrometer. ESI is the most commonly used ionization technique in lipidomics.
5. Mobile phase: The LC system elutes the lipid molecules using a solvent system that includes ammonium acetate in the mobile phases.
6. Column: The CSH C18 column (1.7 μm, 2.1 mm × 100 mm, Waters) is the reverse phase stationary phase in the liquid chromatography system, which separates the lipid molecules based on their physicochemical properties.
7. Software: LipidSearch is used to process and analyze the mass spectra generated by the mass spectrometer [20].

2.2 For GC-MS

1. Reagents: N-hexane n-nadecadecanoate methyl ester (TCI > 99.0%); Chloroform (99.9%); DDH2O (Arium mini); Potassium hydroxide; Methanol; Fatty acid methyl ester mixed standard.
2. Nitrogen generator: used to produce high-purity nitrogen for the ion source of a mass spectrometer.
3. Gas chromatography-mass spectrometry (GC-MS): An Agilent 7890A/5975C gas chromatography-mass spectrometer is used for separation and identification of fatty acids.
4. Gas chromatography column: Agilent DB-WAX capillary column (30 m × 0.25 mm ID × 0.25 μm) is used for separation of fatty acids.
5. Mass spectrometer ion source: used to ionize the sample.
6. Software: The MSD ChemStation software is used to extract chromatographic peak area and retention time [20].

3 Methods

3.1 LC-MS/MS

LC-MS/MS is a widely used lipidomics analysis method that combines liquid chromatography and mass spectrometry to efficiently separate and identify complex lipid molecules. This method involves separating lipid molecules from the sample using liquid chromatography, followed by detection and identification using mass spectrometry. It offers high sensitivity, resolution, and throughput, making it applicable in various lipidomics research areas. Shanghai Applied Protein Technology Co., Ltd. provides detailed information on their LC-MS/MS-based lipidomic methods, which are described below (*see* **Note 1**):

1. Sample preparation [21] (*see* **Note 2**): The MTBE method is used to prepare the samples and extract the lipids. The samples are thawed at 4 °C and mixed with 200 μL of water, 240 μL of methanol, and 800 μL of MTBE containing an internal standard mixture (refer to **Note 1**). The mixture is sonicated at a low temperature (30 min/once, twice) and then centrifuged at 14,000× *g* for 15 min at 10 °C. The upper layer is collected and dried under nitrogen. The lipid extracts are re-dissolved in 200 μL of 90% isopropanol/acetonitrile, centrifuged at 14,000× *g* for 15 min, and finally, 3 μL of the sample is injected. Quality control (QC) samples are prepared by pooling 10 μL of each sample (*see* **Note 3**). These are run at the beginning of the sample queue for column conditioning and every ten injections thereafter to assess inconsistencies that are particularly evident in large batch acquisitions in terms of retention time drifts and variation in ion intensity over time.
2. Chromatographic conditions: Reverse phase chromatography with a CSH C18 column (1.7 μm, 2.1 mm × 100 mm, Waters) is used for LC separation to analyze lipids using LC-MS/MS. The lipid extracts are dissolved in 200 μL of 90% isopropanol/acetonitrile, centrifuged at 14,000× g for 15 min, and 3 μL of the resulting sample is injected. Solvent A consists of acetonitrile–water (6:4, vol/vol) with 0.1% formic acid and 0.1 mM ammonium formate, while solvent B is acetonitrile–isopropanol (1:9, vol/vol) with 0.1% formic acid and 0.1 mM ammonium formate. The initial mobile phase is 30% solvent B at a flow rate of 300 μL/min, held for 2 min, then linearly increased to 100% solvent B in 23 min, followed by equilibrating in 5% solvent B for 10 min. Mass spectra are acquired using Q-Exactive Plus in both positive and negative modes. ESI (Electron spray ionization) parameters are optimized and preset for all measurements, including a source temperature of 300 °C, capillary temperature of 350 °C, ion spray voltage of 3000 V, S-Lens RF Level of 50%, and a scan range of 200–1800 m/z.
3. Mass spectrometry conditions (*see* **Note 4**): ESI is used to detect positive and negative ions. The samples are separated by UHPLC and analyzed by a Q exactive mass spectrometer. The ESI source conditions are as follows: the heater temperature is 300 °C, the sheath gas flow rate is 45 arb, the aux gas flow rate is 15 arb, the sweep gas flow rate is 1 arb, the spray voltage is 3.0 KV, the capillary temperature is 350 °C, the S-Lens RF level is 50%, and MS1 scan ranges from 200 to 1800. The mass charge ratio of lipid molecules and lipid fragments is collected according to the following methods: 10 fragments (MS2scan, HCD) are collected after each full scan. The resolution of MS1 is 70,000 at m/z 200 and that of MS2 is

17,500 at m/z 200. Lipidsearch (Thermo Fisher Scientific, USA) is used to extract and identify the peaks of lipid molecules and internal standard lipid molecules. The main parameters are as follows: precursor tolerance is 5 ppm, product tolerance is 5 ppm, and product ion threshold is 5%.

4. Data processing and analysis (*see* **Note 5**): Lipid species are identified using the LipidSearch software version 4.2 (Thermo Scientific™) to process the raw data and for peak alignment, retention time correction, and extraction peak area. LipidSearch contains data on more than 30 lipid classes, with information on more than 1,700,000 ion fragments. Adducts of +H, +NH4 are selected for positive mode searches, and -H, +CH3COO are selected for negative mode searches since ammonium acetate is used in the mobile phases. For the data extracted from LipidSearch, the ion peak with a value of >50% missing from the group is removed. After normalization and integration using the Perato scaling method, the processed data are imported into SIMPCA-P 16.1 (Umetrics, Umea, Sweden) for a multivariate statistical analysis, including a principal component analysis (PCA), partial least squares discriminant analysis (PLS-DA), and orthogonal partial least squares discriminant analysis (OPLS-DA). Lipids with significant differences are identified based on a combination of statistically significant thresholds of variable influence on projection (VIP > 1) values obtained from the OPLS-DA model (multidimensional statistical analysis) and two-tailed student's t-test (*P*-value <0.05) on the raw data by PASS 16 (https://www.ncss.com/software/pass/) before experiments.
5. Result interpretation: The results are interpreted and comprehensively analyzed based on the experimental design and analysis purpose. Additionally, other experimental data, such as gene expression data and metabolomics data, are combined for systems biology analysis.

3.2 GC-MS

GC-MS is a commonly used analytical method for detecting and identifying volatile compounds. It combines gas chromatography and mass spectrometry techniques to separate and identify complex mixtures. In GC-MS analysis, the sample is first separated by gas chromatography and then ionized and fragmented in the mass spectrometer, resulting in a mass spectrum. By analyzing the mass spectrum, information such as the structure and content of the compound can be determined. The detailed method steps for GC-MS detection of lipidomics are as follows:

1. The samples are thawed in ice, and 60 mg of them is taken into a 2 mL glass centrifuge tube. Then 1 mL of chloroform methanol solution is added. After ultrasonic for 30 min, the

supernatant is taken and 2 ml, 1% sulfuric acid - methanol solution is added. The mixture is then put on the water bath with 80 °C for half-hour methyl esterification. Then 1 ml of n-hexane is added for extraction and 5 ml of water is added for washing. Finally, 500 μl of supernatant of the extract is mixed with 25 μl of internal standard (methyl nonadecanoate) before injection.

2. Standard preparation (*see* **Note 3**): The mixed standard solution of 40 component methyl esterified fatty acids (Sigma-Aldrich) is utilized as reference standards to identify the fatty acids. They calculate the quantity of each methyl fatty ester from calibration curves of standards. Ten mixed standard concentration gradients, ranging from 0.5 mg/L to 1000 mg/L, are used to determine the concentration of each component. In the standard of 40 fatty acid methyl ester, each component accounts for 2% and 4% of the total concentration of two gradients. Specifically, 30 components account for 2% and 10 components account for 4%.
3. GC-MS analysis: The samples are separated on an Agilent DB-WAX capillary column (30 m × 0.25 mm ID×0.25 μm) gas chromatography system. The temperature programming is as follows: the initial temperature is 50 °C and remains as such for 3 min. The temperature is increased at 10 °C/min up to 220 °C and remains there for 20 min. The carrier gas is helium, and the carrier gas velocity is 1.0 mL/min. A QC sample is used for testing and evaluating the stability and repeatability of the system. An Agilent 7890A/5975C gas chromatography-mass spectrometer is used for analysis (*see* **Note 4**). The temperatures of the injection port and transmission line are 280 °C and 250 °C, respectively. Electron bombardment ionization (EI) source, SIM scanning mode, and electron energy of 70 eV are used.
4. Data analysis (*see* **Note 5**): The MSD ChemStation software is used to extract chromatographic peak area and retention time. The content of medium and long-chain fatty acids in the sample is calculated by plotting the curve. The quality control samples are processed together with the biological samples. Detected metabolites in pooled samples with a coefficient of variation (CV) less than 30% are denoted as reproducible measurements. To identify different expressed metabolites, statistical analyses between two sample groups are performed by calculating the Student's t-test p values of metabolites. Metabolites with p values <0.05 are marked as the significantly changed metabolites between sample groups.

4 Notes

1. Laboratory environment and safety: Lipidomics testing must be conducted in a laboratory environment, and it is necessary to ensure the cleanliness and safety of the laboratory. Laboratory personnel should follow safety regulations to ensure their health and safety.
2. Collection and processing of samples: Standard operating procedures should be followed for the collection and processing of samples to ensure the accuracy and reproducibility of lipidomics testing. Sample harvesting should be done under the same conditions to avoid environmental perturbations, and samples should be snap-frozen in liquid nitrogen for enzyme quenching. Stored samples should be kept at low temperatures, and factors that may affect sample quality, such as oxidation and contamination, should be avoided.
3. Internal standards preparation: To ensure accuracy, it is recommended to use at least three evenly distributed internal standards throughout the retention time dimension. Calibration standards should also be prepared for method validation. Calibration curves can be constructed using nominal concentrations of lipid calibrators (X-axis) and MS peak area ratios of the lipid standard. Standard samples are crucial for lipidomics testing, and it is important to prepare and use them according to standard operating procedures.
4. Instrument selection and calibration: Lipidomics testing requires high sensitivity and high-resolution instruments, such as mass spectrometers and chromatographs. Before using these instruments, they should be calibrated and validated to ensure their accuracy and reliability.
5. Data processing and analysis: Lipidomics testing produces a large amount of data, and professional data processing and analysis software should be used to process and analyze the data. Quality control and filtering should be performed before processing and analyzing the data to ensure its accuracy and reliability.

References

1. Lin Z, Liu J, Kang R et al (2021) Lipid metabolism in Ferroptosis. Adv Biol (Weinh) 5: e2100396
2. Xie Y, Hou W, Song X et al (2016) Ferroptosis: process and function. Cell Death Differ 23: 369–379
3. Thiam AR, Farese RV Jr, Walther TC (2013) The biophysics and cell biology of lipid droplets. Nat Rev Mol Cell Biol 14:775–786
4. Lionetti L, Mollica MP, Lombardi A et al (2009) From chronic overnutrition to insulin resistance: the role of fat-storing capacity and inflammation. Nutr Metab Cardiovasc Dis NMCD 19:146–152
5. Tang D, Kroemer G, Kang R (2023) Ferroptosis in immunostimulation and immunosuppression. Immunol Rev doi: 10.1111/imr.13235. Epub ahead of print.

6. Harayama T, Riezman H (2018) Understanding the diversity of membrane lipid composition. Nat Rev Mol Cell Biol 19:281–296
7. Fahy E, Sud M, Cotter D et al (2007) LIPID MAPS online tools for lipid research. Nucleic Acids Res 35:W606–W612
8. Yang WS, Kim KJ, Gaschler MM et al (2016) Peroxidation of polyunsaturated fatty acids by lipoxygenases drives ferroptosis. Proc Natl Acad Sci U S A 113:E4966–E4975
9. Kagan VE, Mao G, Qu F et al (2017) Oxidized arachidonic and adrenic PEs navigate cells to ferroptosis. Nat Chem Biol 13:81–90
10. Yuan H, Li X, Zhang X et al (2016) Identification of ACSL4 as a biomarker and contributor of ferroptosis. Biochem Biophys Res Commun 478:1338–1343
11. Lin Z, Liu J, Long F et al (2022) The lipid flippase SLC47A1 blocks metabolic vulnerability to ferroptosis. Nat Commun 13:7965
12. Magtanong L, Ko PJ, To M et al (2019) Exogenous monounsaturated fatty acids promote a Ferroptosis-resistant cell state. Cell Chem Biol 26:420–432. e9, 420
13. Liu Y, Wang Y, Lin Z et al (2023) SLC25A22 as a key mitochondrial transporter against ferroptosis by producing GSH and MUFAs. Antioxid Redox Signal
14. Ni Z, Wölk M, Jukes G et al (2023) Guiding the choice of informatics software and tools for lipidomics research applications. Nat Methods 20:193–204
15. Alseekh S, Aharoni A, Brotman Y et al (2021) Mass spectrometry-based metabolomics: a guide for annotation, quantification and best reporting practices. Nat Methods 18:747–756
16. Shevchenko A, Simons K (2010) Lipidomics: coming to grips with lipid diversity. Nat Rev Mol Cell Biol 11:593–598
17. Gross JH (2017) Mass spectrometry: a textbook. Springer, Heidelberg
18. Cui L, Lu H, Lee YH (2018) Challenges and emergent solutions for LC-MS/MS based untargeted metabolomics in diseases. Mass Spectrom Rev 37:772–792
19. Drexler DM, Reily MD, Shipkova PA (2011) Advances in mass spectrometry applied to pharmaceutical metabolomics. Anal Bioanal Chem 399:2645–2653
20. Fiehn O (2016) Metabolomics by gas chromatography-mass spectrometry: combined targeted and untargeted profiling. Curr Protocols Mol Biol 114:30.4.1–30.4.2
21. Tascon M, Gómez-Ríos GA (2023) Direct sample preparation mass spectrometry analysis: toward greener analytical methods. Green Approach Chem Analysis Elsevier:127–149

Chapter 14

In-Cell Western Assay in Ferroptosis

Jiayi Wang and Jingbo Li

Abstract

Ferroptosis is a regulated form of cell death caused by the excessive accumulation of iron-dependent lipid peroxidation. It has been implicated in various pathological processes and diseases, and its modulation involves multiple proteins associated with iron and lipid metabolism. To better understand these mechanisms and monitor the ferroptosis process, there is a need for reliable and high-throughput methods to evaluate variations in protein expression levels. In-Cell Western assays provide a simple and rapid assay method for detecting biomarkers and signaling proteins in whole cells using antibodies. This assay involves seeding cells in microtiter plates, followed by fixation/permeabilization and subsequent labeling with primary antibodies and infrared-conjugated secondary antibodies. In this chapter, we introduce the protocol for the In-Cell Western assay for detecting intracellular proteins during ferroptosis.

Key words Ferroptosis, In-Cell Western assay, Protein expression, High-throughput

1 Introduction

Cell death is a fundamental biological process that regulates cell fate, immune response, and tissue homeostasis. It can be classified into regulated cell death (RCD) or accidental cell death (ACD) [1]. In contrast to ACD, which is an uncontrolled and passive process, RCD is explicitly executed by multiple genes and proteins [2]. Ferroptosis is a form of iron- and lipotoxicity-dependent RCD that is distinct from other types of RCD, such as apoptosis and necroptosis [3, 4]. Proteins serve as important regulatory factors that regulate ferroptosis by directly or indirectly targeting lipid peroxidation or iron metabolism [5]. Additionally, characteristic molecular alterations can be used as specific markers for the occurrence of ferroptosis [6]. Therefore, detecting ferroptosis-associated protein levels could help reveal the mechanisms of ferroptosis and provide novel insights for disease therapy [7–11].

Several core regulatory molecules of ferroptosis have been identified, including glutathione peroxidase 4 (GPX4) [12], solute carrier family 7 member 11 (SLC7A11) [3, 13–15],

Guido Kroemer and Daolin Tang (eds.), *Ferroptosis: Methods and Protocols*, Methods in Molecular Biology, vol. 2712, https://doi.org/10.1007/978-1-0716-3433-2_14,

apoptosis-inducing factor mitochondria-associated 2 (AIFM2) [16, 17], acyl-CoA synthetase long-chain family member 4 (ACSL4) [18, 19], the nuclear factor, erythroid 2-like 2 (NFE2L2, also known as NRF2) [20], and dihydroorotate dehydrogenase (DHODH) [21]. During ferroptosis, GPX4 is depleted and downregulated, while ACSL4 expression is upregulated at the transcriptional level [19]. Interestingly, the change in SLC7A11 expression was opposite in diverse tumor models following treatment with erastin, a well-known ferroptosis inducer [22]. Additionally, ferroptosis has been proposed as an autophagy-dependent cell death process, indicating that components of autophagy are required to initiate or accelerate the main events of ferroptotic cell death, such as lipid peroxidation and excess accumulation of free iron [23–25]. Selective autophagy, including ferritinophagy [26], lipophagy [27], and clockophagy [28], has been found to promote ferroptosis by degrading anti-ferroptotic regulators in a context-dependent manner [24, 29–33]. Therefore, detecting the expression of proteins involved in autophagic degradation may assist in understanding the processes and functions of ferroptosis.

There are several approaches for assessing cellular protein expression levels. The most commonly used method is western blotting (WB), which directly analyzes cell lysates for the levels of target proteins. However, WB requires several steps and limits the number of samples that can be analyzed in each experiment. Although Enzyme-Linked Immunosorbent Assays (ELISA) are a representative high-throughput method for measuring secreted proteins, they are rarely used for detecting intracellular proteins. Quantitative reverse transcription polymerase chain reaction (qRT-PCR) analysis can detect gene transcript levels as an approximation of protein expression levels. However, post-transcriptional and translational modifications can influence protein expression, suggesting that transcript levels do not fully represent protein levels.

In this chapter, we will introduce the In-Cell Western (ICW) assay as a potential method for detecting changes in protein expression directly in cultured mammalian cells, including in the context of ferroptosis. This method combines the replicability and throughput of ELISA with the specificity of WB, and can quantify multiple target proteins simultaneously in a cellular culture system [34, 35]. The ICW assay has been widely applied in diverse research areas, such as the quantification of viral load [36–38], chemical library screening [39, 40], protein phosphorylation [34, 41], glycoprotein analysis [42], and screening of monoclonal antibody clones [43]. Given its features and advantages, ICW should be also considered as a means of assessing the expression of intracellular proteins involved in ferroptosis, which may lead to new insights and potential therapeutic targets for this important biological process.

2 Materials

1. PANC1 or HT1080 cells.
2. Dulbecco's modified Eagle medium (DMEM).
3. Black-walled 96- or 384-well cell culture plates.
4. Fetal Bovine Serum (FBS).
5. 10× Phosphate Buffered Saline (PBS): To prepare 1 L add 80 g sodium chloride (NaCl), 2 g potassium chloride (KCl), 14.4 g sodium phosphate, dibasic (Na_2HPO_4) and 2.4 g potassium phosphate, monobasic (KH_2PO_4) to 1 L dH_2O. Adjust pH to 7.4.
6. Pipette tips.
7. Protease and phosphatase inhibitors.
8. Cold room.
9. Plate shaker.
10. Dimethyl sulfoxide (DMSO).
11. Fixing solution: 4% formaldehyde in PBS or 100% methanol.
12. Permeabilization solution: 0.1% Triton X-100 in PBS.
13. Blocking buffer: To prepare 25 mL of blocking buffer, add 2.5 mL of 10× PBS, 1.25 mL of normal serum from the same species as the secondary antibody, and 21.25 mL of dH2O. Mix well and add 75 μL of Triton X-100 while stirring.
14. Primary antibody: anti-GPX4 antibody (Abcam, #ab125066) and anti-ACSL4 antibody (Abcam, #ab240135).
15. Fluorochrome-conjugated secondary antibody: Anti-rabbit IgG (H + L) (DyLight™ 680 Conjugate), Cell Signaling Technology, #5366; Anti-rabbit IgG (H + L) (DyLight™ 800 4X PEG Conjugate), Cell Signaling Technology, #5151.
16. Antibody dilution buffer: To prepare 40 mL of the solution, add 4 mL of 10× PBS to 36 mL of dH2O and mix well. Add 0.4 g of BSA and mix until dissolved. While stirring, add 120 μL of 100% Triton X-100.
17. LI-COR Odyssey infrared imaging system.

3 Methods

ICW is a quantitative immunofluorescence (IF) assay performed in microplates that can directly quantify intracellular protein expression based on fluorescence intensity [44]. The assay involves seeding target cells and treating them with ferroptosis inducers (e.g., erastin or RSL3) in microplates, followed by fixation and permeabilization to protect cell structures and allow the antibody to enter the cellular and nuclear membranes. Blocking buffer is then added to bind nonspecific binding sites in the cells and fill spaces in the wells that are not occupied by cells. The target protein is recognized using a specific primary antibody, followed by an infrared-labeled secondary antibody that binds to the primary antibody. Finally, the samples labeled with IF are scanned, quantified, and analyzed by a two-channel near-infrared scanner.

1. Seed PANC1 or HT1080 cells at an appropriate density in multi-well plates (96 or 384 wells) and incubate overnight at 37 °C (*see* **Notes 1** and **2**).
2. Prepare all reagents on the day of the experiment.
3. Treat the cells with RSL3 (0.5 μM) or erastin (5 μM) and incubate for 6–24 h.
4. Fix cells with 50 μL 4% formaldehyde in 1× PBS or 100% methanol at room temperature for 15 min (*see* **Note 3**).
5. Remove the fixation solution and rinse the plate three times with room temperature 1× PBS for 5 min each time (*see* **Note 4**).
6. Permeabilize the cells by adding 0.1% Triton X-100 and gently rotating the plate for 5 min at room temperature. Discard the permeabilization solution and repeat the wash 3 times with 1× PBS (*see* **Note 4**).
7. Block cells by adding blocking buffer to each well and incubate for 1 h at room temperature with gentle shaking (*see* **Note 5**).
8. Discard the blocking solution and add diluted primary antibody (e.g., anti-GPX4 antibody 1:1000; *see* **Notes 6** and 7). Cover the plate and incubate overnight at 4 °C.
9. Wash the plate 3 times with 1× PBS for 5 min each time (*see* **Note 4**).
10. Add diluted fluorochrome-conjugated secondary antibody (1: 10–1:40) to each well and incubate for 1 h in the dark at room temperature (*see* **Note 8**).
11. Wash the plate 3 times with 1× PBS for 5 min each time (*see* **Note 4**).
12. Scan the dry plate on the LI-COR® Odyssey™ or an equivalent imaging system according to the manufacturer's instructions (*see* **Note 9**).
13. Use Image Studio software or equivalent to quantify the fluorescent signals (*see* **Note 10**).

4 Notes

1. To minimize background fluorescence and false signals from scattered light, use black-walled multi-well plates.
2. Let newly seeded plates stand at room temperature for 1 h before placing them in a 37 °C incubator to avoid edge effects from uneven distribution of cells in single wells [45].
3. When adding liquid, be careful to pour it down the sides of the wells to avoid detaching the well bottom cells.
4. Pour the liquid into a waste receptacle and cover the well plate with a clean paper towel, invert it, and shake it sharply 2–3 times to empty the wells. Avoid touching the inside of the wells with aspirators or pipettes as this may leave artifacts and affect subsequent quantification.
5. Blocking is an important step in the ICW assay to minimize the non-specific binding of primary and secondary antibodies to non-target sites in cells and wells. There is no single blocking buffer that is optimal for each target, so it is important to choose appropriate blocking conditions for target proteins and cell lines. If it is difficult to detect target proteins, the blocking agent and blocking time should be adjusted.
6. For double labeling, prepare a cocktail of rabbit and mouse primary antibodies at their appropriate dilution as suggested in the product instructions.
7. Primary antibodies are essential for the ICW assay and should be validated by immunocytochemistry, immunohistochemistry, or IF. The performance and specificity of the antibodies should be verified through immunocytochemistry and western blotting [34]. Ideally, the antibodies should be pre-adsorbed and affinity-purified with serum from other animal species or other proteins to remove cross-reactivity. In addition, the working concentration of the primary antibody is variable, and the antibodies should be diluted as suggested by the manufacturers.
8. For double labeling, prepare a cocktail of fluorochrome-conjugated anti-rabbit and anti-mouse secondary antibodies at the appropriate dilutions according to the manufacturer's instructions.
9. Wipe the bottom of the plate before scanning.
10. Normalization is necessary to account for unavoidable cell number differences. Some methods that can be used for ICW normalization rely on the detection of either a cell stain or a constitutively expressed protein, and the choice of

normalization method depends on which strategy is appropriate for the experimental conditions.

References

1. Galluzzi L, Vitale I, Aaronson SA et al (2018) Molecular mechanisms of cell death: recommendations of the nomenclature committee on cell death 2018. Cell Death Differ 25: 486–541
2. Tang D, Kang R, Berghe TV et al (2019) The molecular machinery of regulated cell death. Cell Res 29:347–364
3. Dixon SJ, Lemberg KM, Lamprecht MR et al (2012) Ferroptosis: an iron-dependent form of nonapoptotic cell death. Cell 149:1060–1072
4. Xie Y, Hou W, Song X et al (2016) Ferroptosis: process and function. Cell Death Differ 23: 369–379
5. Liu J, Kang R, Tang D (2022) Signaling pathways and defense mechanisms of ferroptosis. FEBS J 289:7038–7050
6. Chen X, Comish PB, Tang D et al (2021) Characteristics and biomarkers of Ferroptosis. Front Cell Dev Biol 9:637162
7. Chen X, Kang R, Kroemer G et al (2021) Broadening horizons: the role of ferroptosis in cancer. Nat Rev Clin Oncol 18:280–296
8. Tang D, Chen X, Kang R et al (2021) Ferroptosis: molecular mechanisms and health implications. Cell Res 31:107–125
9. Stockwell BR, Friedmann Angeli JP, Bayir H et al (2017) Ferroptosis: a regulated cell death nexus linking metabolism, redox biology, and disease. Cell 171:273–285
10. Tang D, Kroemer G, Kang R (2023) Ferroptosis in hepatocellular carcinoma: from bench to bedside. Hepatology. Publish Ahead of Print
11. Chen X, Kang R, Kroemer G et al (2021) Targeting ferroptosis in pancreatic cancer: a double-edged sword. Trends Cancer 7:891–901
12. Yang WS, SriRamaratnam R, Welsch ME et al (2014) Regulation of ferroptotic cancer cell death by GPX4. Cell 156:317–331
13. Jiang L, Kon N, Li T et al (2015) Ferroptosis as a p53-mediated activity during tumour suppression. Nature 520:57–62
14. Zhang Y, Shi J, Liu X et al (2018) BAP1 links metabolic regulation of ferroptosis to tumour suppression. Nat Cell Biol 20:1181–1192
15. Song X, Zhu S, Chen P, et al. (2018) AMPK-mediated BECN1 phosphorylation promotes Ferroptosis by directly blocking system X(c)(−) activity. Curr Biol 28: 2388–99 e5, 2388
16. Doll S, Freitas FP, Shah R et al (2019) FSP1 is a glutathione-independent ferroptosis suppressor. Nature 575:693–698
17. Dai E, Zhang W, Cong D et al (2020) AIFM2 blocks ferroptosis independent of ubiquinol metabolism. Biochem Biophys Res Commun 523:966–971
18. Doll S, Proneth B, Tyurina YY et al (2017) ACSL4 dictates ferroptosis sensitivity by shaping cellular lipid composition. Nat Chem Biol 13:91–98
19. Yuan H, Li X, Zhang X et al (2016) Identification of ACSL4 as a biomarker and contributor of ferroptosis. Biochem Biophys Res Commun 478:1338–1343
20. Sun X, Ou Z, Chen R et al (2016) Activation of the p62-Keap1-NRF2 pathway protects against ferroptosis in hepatocellular carcinoma cells. Hepatology 63:173–184
21. Mao C, Liu X, Zhang Y et al (2021) Author correction: DHODH-mediated ferroptosis defence is a targetable vulnerability in cancer. Nature 596:E13
22. Chen X, Li J, Kang R et al (2021) Ferroptosis: machinery and regulation. Autophagy 17: 2054–2081
23. Chen F, Cai X, Kang R et al (2023) Autophagy-dependent Ferroptosis in cancer. Antioxid Redox Signal
24. Li J, Liu J, Xu Y et al (2021) Tumor heterogeneity in autophagy-dependent ferroptosis. Autophagy 17:3361–3374
25. Liu J, Kuang F, Kroemer G et al (2020) Autophagy-dependent Ferroptosis: machinery and regulation. Cell Chem Biol 27:420–435
26. Hou W, Xie Y, Song X et al (2016) Autophagy promotes ferroptosis by degradation of ferritin. Autophagy 12:1425–1428
27. Bai Y, Meng L, Han L et al (2019) Lipid storage and lipophagy regulates ferroptosis. Biochem Biophys Res Commun 508:997–1003
28. Yang M, Chen P, Liu J et al (2019) Clockophagy is a novel selective autophagy process favoring ferroptosis. Sci Adv 5:eaaw2238
29. Zhou B, Liu J, Kang R et al (2020) Ferroptosis is a type of autophagy-dependent cell death. Semin Cancer Biol 66:89–100
30. Liu J, Liu Y, Wang Y et al (2023) TMEM164 is a new determinant of autophagy-dependent ferroptosis. Autophagy 19:945–956

31. Chen X, Song X, Li J et al (2022) Identification of HPCAL1 as a specific autophagy receptor involved in ferroptosis. Autophagy:1–21
32. Kang R, Zhu S, Zeh HJ et al (2018) BECN1 is a new driver of ferroptosis. Autophagy 14: 2173–2175
33. Chen X, Song X, Li J et al (2023) Identification of HPCAL1 as a specific autophagy receptor involved in ferroptosis. Autophagy 19:54–74
34. Aguilar HN, Zielnik B, Tracey CN et al (2010) Quantification of rapid myosin regulatory light chain phosphorylation using high-throughput in-cell Western assays: comparison to Western immunoblots. PLoS One 5:e9965
35. Coevoets R, Arican S, Hoogeveen-Westerveld M et al (2009) A reliable cell-based assay for testing unclassified TSC2 gene variants. Eur J Hum Genet EJHG 17:301–310
36. Wan Y, Zhou Z, Yang Y et al (2010) Application of an in-cell Western assay for measurement of influenza a virus replication. J Virol Methods 169:359–364
37. DuShane JK, Wilczek MP, Crocker MA et al (2019) High-throughput characterization of viral and cellular protein expression patterns during JC polyomavirus infection. Front Microbiol 10:783
38. Ma HW, Ye W, Chen HS et al (2017) In-cell Western assays to evaluate Hantaan virus replication as a novel approach to screen antiviral molecules and detect neutralizing antibody titers. Front Cell Infect Microbiol 7:269
39. Hoffman GR, Moerke NJ, Hsia M et al (2010) A high-throughput, cell-based screening method for siRNA and small molecule inhibitors of mTORC1 signaling using the in cell Western technique. Assay Drug Dev Technol 8:186–199
40. Schnaiter S, Fürst B, Neu J et al (2014) Screening for MAPK modulators using an in-cell western assay. Methods Mol Biol (Clifton, NJ) 1120:121–129
41. Wong SK (2004) A 384-well cell-based phospho-ERK assay for dopamine D2 and D3 receptors. Anal Biochem 333:265–272
42. McInerney MP, Pan Y, Short JL et al (2017) Development and validation of an in-cell Western for quantifying P-glycoprotein expression in human brain microvascular endothelial (hCMEC/D3) cells. J Pharm Sci 106:2614–2624
43. Daftarian MP, Vosoughi A, Lemmon V (2014) Gene-based vaccination and screening methods to develop monoclonal antibodies. Methods Mol Biol (Clifton, NJ) 1121:337–346
44. Ruiz-Del-Yerro E, Garcia-Jimenez I, Mamchaoui K et al (2018) Myoblots: dystrophin quantification by in-cell western assay for a streamlined development of Duchenne muscular dystrophy (DMD) treatments. Neuropathol Appl Neurobiol 44:463–473
45. Lundholt BK, Scudder KM, Pagliaro L (2003) A simple technique for reducing edge effect in cell-based assays. J Biomol Screen 8:566–570

Chapter 15

Flow Cytometric Analysis of Regulated Cell Death

Siyuan Huang and Ling Zeng

Abstract

Cell death is a crucial and fundamental process in the biology of all living organisms and plays an essential role in developmental, cellular, and molecular biology. With a deeper understanding of the mechanisms of different types of cell death, quantitative experiments are becoming increasingly necessary to analyze the dynamic changes and coordinate physiological processes. Flow cytometry is the most widely used method for detecting and quantifying cell processes in mammalian cells, providing a comprehensive and high-throughput approach, even at the single-cell level. This chapter provides a brief overview of guidelines for performing flow cytometry in the detection of regulated cell deaths, including apoptosis, autophagy, ferroptosis, pyroptosis, and immunogenic cell death.

Key words Apoptosis, Autophagy, Ferroptosis, Flow cytometry, Immunogenic cell death, Pyroptosis

1 Introduction

As recognized by the Cell Death Nomenclature Committee, cell death is broadly divided into two categories: accidental cell death and regulated cell death (RCD) [1]. RCD is a programmed death controlled by genes and proteins and occurs through various cellular signaling pathways [1]. In recent years, RCD has garnered increasing attention due to its predictability and controllability, leading to a better understanding of the various subclasses of RCD, including apoptosis, pyroptosis, necroptosis, and ferroptosis, distinguished by their unique morphological characteristics, gene expression diversity, and physiological and biochemical alterations [2]. It's also worth noting that autophagy, a lysosomal degradation pathway, can have either inhibitory or promoting effects on cell death [3]. However, there is limited research into the methodologies for characterizing these biological features.

For the past three decades, flow cytometry has been widely used to detect and quantify RCD in mammalian cells [4]. The process involves the preparation of single-cell suspensions from cultured cells, Red blood cell (RBC)-lysed whole blood cells, or

Guido Kroemer and Daolin Tang (eds.), *Ferroptosis: Methods and Protocols*, Methods in Molecular Biology, vol. 2712, https://doi.org/10.1007/978-1-0716-3433-2_15,

dissociated tissue-derived cells, followed by the addition of fluorescein-conjugated primary antibodies to the cell suspension. These antibodies label specific cell surface or intracellular molecule antigens through antigen-antibody interaction, which is then analyzed by flow cytometry. The cell viability can be assessed by differences in forward scatter (FSC), side scatter (SSC), and mean fluorescence intensity (MFI) of the specific fluorescein [5]. Compared to microscopy-based techniques like immunohistochemistry and immunocytochemistry, which require cell lysis procedures, flow cytometry assays are quantitative, rapid, high-throughput, and most notably, capable of detecting RCD at the single-cell level [6]. Additionally, flow cytometry assays offer a more comprehensive method for characterizing the entire cell death process and sorting live cells for further experimentation [7].

Investigating the processes and features of different subclasses of RCD is key to developing strategies for manipulating cell survival, and can also offer valuable insights into the treatment of cancer and immune diseases. This chapter will provide an overview of flow cytometry assays for different subclasses of RCD.

2 Materials

1. Cells of interest: such as PANC1 and HT1080 cells.
2. Buffer solution: such as eBioscience™ Flow Cytometry Staining Buffer (# 00–4222-26, ThermoFisher).
3. Fluorescently labeled antibodies or probes: such as Annexin V-FITC/PI and C11-BODIPY 581/591.
4. Isotype controls: These are used to measure nonspecific binding of the antibodies to the cells, and to ensure that the fluorescence signal is specific to the cell surface markers being analyzed.
5. Flow cytometer: such as BD Accuri C6.
6. Sample tubes: such as MTC Bio FlowTubes™ Flow Cytometry/FACS Tubes (#50–187-1142. ThermoFisher).
7. Sheath fluid: such as BD FACS™ Sheath Fluid (#342003).
8. Calibration beads: Used to calibrate the instrument and ensure accurate measurements.
9. Data analysis software: such as BD FACSDiva™ Software.

3 Methods

3.1 Flow Cytometry Assay in Apoptosis

Apoptosis is a common form of programmed cell death and occurs as a physiological or pathological response to adapt to the environment. It is distinct from necrosis as it is an active process that is tightly regulated by genes, such as the BCL-2 family, the caspase family, proto-oncogenes, and anti-oncogenes [8].

Apoptosis plays a crucial role in maintaining homeostasis and can be activated through either extrinsic or intrinsic pathways. In the extrinsic pathway, the binding of FAS-ligand to the FAS receptor on the cell surface triggers the activation of caspase 8, which then activates the caspase-3 cascade. In the intrinsic pathway, the formation of the apoptosome, consisting of cytochrome c released from mitochondria and apoptotic protease activating factor-1 (APAF-1), activates caspase-9 and ultimately caspase-3 to initiate the execution of the apoptosis program [2, 9, 10].

Flow cytometry assays use primary antibodies that bind to specific antigens (or markers) to detect apoptosis. The primary markers of apoptosis include caspase activation, degradation of membrane permeability, and the flipping of phosphatidylserine (PS). The major flow cytometry assays for apoptosis include Annexin-V and PI (propidium iodide) co-dyeing assay, fluorogenic caspase assay/FLICA (fluorochrome-labeled inhibitors of caspases) assay, and terminal deoxynucleotidyl transferase dUTP nick end labeling (TUNEL) assay, which determine the early, mid-term, and late phases of apoptosis, respectively. To ensure a successful flow cytometry experiment, it's essential to pay attention to some of the most basic factors (*see* **Notes 1–8**).

3.1.1 Annexin V-PI Co-dyeing Assay

A modified one-step Annexin V-PI co-dyeing assay protocol, based on typical cell culture practices, is described in detail below. This protocol is based on the Annexin V-FITC/PI Apoptosis Kit available from Elabsciences (https://www.elabscience.cn/p-annexin_v_fitc_pi_apoptosis_kit-203045.html).

1. Induce apoptosis or treat the cells using the desired method.
2. Collect and quantify the cells. Divide the cells into a test group, a control group, and a negative control group. Adjust the cell count to 1–5 × 10^5 cells per group. Centrifuge the cells at 350 *g* for 5 min and discard the supernatant.
3. Wash the cell pellet with phosphate-buffered saline (PBS) once and centrifuge again at 350 *g* for 5 min. Discard the supernatant.
4. Resuspend the cell pellet in 500 μL of the diluted 1 × Annexin V Binding Buffer.
5. Add 5 μL of Annexin V dye and 5 μL of PI dye.
6. Gently vortex the solution and incubate at 37 °C in the dark for 10–15 min.
7. Analyze the sample using a flow cytometer within 1 h.

3.1.2 Fluoregenic Caspase Assay

A modified fluorogenic caspase assay protocol, based on typical cell culture practices, is described in detail below. This protocol is based on the Caspase Assay Kit for Flow Cytometry available from Abcam (https://www.abcam.cn/caspase-assay-kit-flow-cytometry-ab273272.html).

1. Induce apoptosis or treat the cells using the desired method.
2. Simultaneously incubate a control culture without any treatment.
3. Count the cells and pellet 1×10^5 cells.
4. Resuspend the cells in 0.5 mL of D2R Incubation Buffer.
5. Add 4 μL of the 1 M dithiothreitol (DTT, 8 mM final concentration).
6. Add 1 μL of the D2R Reagent.
7. Incubate the solution at 37 °C in the dark for 10–20 min.
8. Analyze the cells using a flow cytometer.

3.1.3 FLICA Assay

A modified FLICA assay protocol, based on typical cell culture practices, is described in detail below. This protocol is based on the FLICA Assay Kit from Invitrogen (https://www.thermofisher.cn/order/catalog/product/V35117).

1. Induce apoptosis or treat the cells using the desired method.
2. Harvest the cells and resuspend them to a concentration of 1×10^6 cells/mL in the culture media.
3. Transfer 300 μL of the cell suspension to flow tubes.
4. Add 10 μL of the 30× FLICA working solution directly to the 300 μL cell suspension.
5. Mix the cells by gently flicking the tubes.
6. Incubate the mixture for 60 min at 37 °C and 5% CO_2 in the dark. Make sure to mix the tubes twice during the incubation period to minimize cell settling.
7. Add 2 mL of PBS to each tube.
8. Pellet the cells by centrifuging at 350 *g* for 5 min, discard the supernatant, and resuspend the cells in 1 mL of PBS. Repeat the centrifugation process once.
9. Discard the supernatant and resuspend the cell pellet in 400 μL of PBS.
10. Add 2 μL of the propidium iodide stain to the cell suspension and incubate it for 10 min on ice in the dark.
11. Analyze the samples on a flow cytometer with an excitation wavelength of 488 nm and emission wavelengths of 530/670 nm.

3.1.4 TUNEL Assay

A modified TUNEL flow cytometry assay protocol from Elabsciences (https://www.elabscience.cn/p-one_step_tunel_flow_cytometry_apoptosis_kit_green_fitc_-429094.html) is described below with greater detail and clarity:

1. Induce apoptosis in cells using the desired method.
2. Collect and count the cells, dividing them into three groups: a test group, a control group, and a negative control group. Adjust the cell count to 1 million cells per group. Centrifuge at 350 *g* for 5 min and discard the supernatant.
3. Resuspend the cells in 200 μL of PBS solution containing 1% bovine serum albumin (BSA). Fix the cells with 200 μL of the fixation buffer at room temperature for 60 min, stirring the mixture every 15 min.
4. Centrifuge the cells with 1 mL of PBS at 600 *g* for 5 min and discard the supernatant.
5. Add 200 μL of the permeabilization buffer (pre-warmed at 4 °C), thaw on ice and permeabilize for 10 min.
6. Stop the permeabilization by adding 1–2 mL of PBS and gently stirring. Centrifuge at 600 *g* for 5 min and discard the supernatant.
7. Add 200 μL of the TdT Equilibration Buffer, gently stir, and incubate at 37 °C for 15 min.
8. Centrifuge at 600 g for 5 min and discard the supernatant. Collect the pellet.
9. Prepare the working buffer and add 100 μL of the working buffer per sample (Table 1).
10. Incubate the samples in the dark at 37 °C for 60 min, stirring every 20 min.
11. Add 1 mL of the Stop Solution, gently mix and let stew at room temperature for 5 min. Add 300–500 μL of PBS. Centrifuge at 600 *g* for 5 min and discard the supernatant.
12. Resuspend the cell pellet with 200 μL of PBS and analyze using a flow cytometer as soon as possible.

3.2 Flow Cytometry Assay in Autophagy

Autophagy is a crucial process in eukaryotes that promotes protein or organelle turnover and maintains homeostasis. It was first discovered in 1966 by Duve and Wattiaux, who identified it as a component of lysosomes [11]. A decade later, Mortimore and Schworer discovered its vital role in maintaining hepatocyte homeostasis during starvation [12].

Autophagy is typically triggered by pathogens, starvation, reactive oxygen species (ROS), and endoplasmic reticulum stress, which leads to the activation of autophagy-related genes (ATG) and the formation of autophagic vesicles [13]. Microtubule-associated

Table 1
Working buffer recipe

	Sample	Negative control
TdT equilibration buffer	70 μL	80 μL
Labeling solution	20 μL	20 μL
TdT enzyme	10 μL	0 μL

protein 1A/1B-light chain 3 (LC3) is a ubiquitous, soluble protein with a molecular weight of around 17 kDa found in mammalian tissues and cell cultures. During autophagy, LC3-I is conjugated to phosphatidylethanolamine to form LC3-phosphatidylethanolamine conjugate (LC3-II), which associates with autophagosomal membranes as they engulf cytoplasmic components, including proteins and organelles. These autophagosomes then fuse with lysosomes, where their cargo, including misfolded proteins and damaged organelles, is degraded by lysosomal enzymes to form autolysosomes [14, 15].

As a result, the markers of autophagy mainly focus on acidic vesicles, LC3-II, and autophagy flux. When performing flow cytometry, the extent of autophagy can be determined by the acridine orange (AO) assay, which labels acidic vacuolar organelles (primarily late autophagosomes) and reflects autophagy through fluorescence. The LC3-II assay can be used to label autophagosomes and detect autophagy initiation, while the LysoTracker assay can label autolysosomes and determine autophagy flux dynamics.

3.2.1 AO Assay

A modified AO assay protocol from Sigma-Aldrich (https://www.sigmaaldrich.cn/CN/zh/product/sigma/a6014) is described as follows:

1. Induce autophagy or treat the cells using the desired method.
2. Incubate the cells with 2.7 μM (1 μg/mL) AO in a complete culture medium at room temperature for 15 min.
3. Harvest the cells from the plate using trypsin-ethylenediaminetetraacetic acid (EDTA) for 1–2 min. Centrifuge the cells at 350 *g* for 5 min and discard the supernatant.
4. Resuspend the cell pellet in 1 mL of a complete culture medium free of phenol red.
5. Analyze the sample on a flow cytometer equipped with imaging capabilities.

Note: The treatment may increase the size of the cells, leading to an increased uptake of AO and false-positive or false-negative results. To avoid this, it is recommended to set a threshold based on the proportion of red to green fluorescence intensity [16]

3.2.2 LC3-II Assay

A modified LC3-II assay protocol (based on typical cell culture) from Abcam (https://www.abcam.cn/lc3b-antibody-n-terminal-ab229327.html) is described as follows:

1. Induce autophagy or treat cells using the desired method.
2. Remove the cells from the plate using trypsin-EDTA for 1–2 min and centrifuge at 350 *g* for 5 min. Discard the supernatant.
3. Resuspend the cell pellet in 1 mL of PBS.
4. Fix the cells with 4% paraformaldehyde for 10–15 min.
5. Wash the cells with 1 mL of PBS, centrifuge at 350 *g* for 5 min and discard the supernatant. Repeat the washing step once.
6. Resuspend the cell pellets in 1 mL of PBS and add 1 mL of pre-cold 100% methanol. Gently vortex and incubate at −20 °C for 10 min.
7. Wash the cells with 2 mL of PBS, centrifuge at 350 *g* for 5 min, discard the supernatant, and resuspend the pellet in 100 μL of PBS. Repeat the washing step once.
8. Add anti-LC3B antibody (diluted 1:50) and incubate at 2–8 °C for 30 min in the dark.
9. Wash the cells with PBS, centrifuge at 350 *g* for 5 min, and discard the supernatant. Repeat the washing step twice.
10. Resuspend the cell pellets in 400 μL of PBS.
11. Determine the mean fluorescent intensity on a flow cytometer using an Argon ion laser (488 nm) and a 525/30 bandpass filter.

3.2.3 LysoTracker Assay

A modified LysoTracker Green assay protocol, as described by Invitrogen (https://www.thermofisher.cn/order/catalog/product/L7526?SID=srch-hj-L7526), is outlined as follows:

1. Induce autophagy or treat cells using the desired method.
2. Dilute the 1 mM stock solution of the probe to a final working concentration of 50 nM in the growth medium or buffer of choice. It's important to keep the dye concentration as low as possible to avoid potential artifacts from overloading.
3. For adherent cells, grow them on coverslips inside a Petri dish with the appropriate culture medium. Once the cells have reached the desired confluence, remove the medium from the dish and add pre-warmed (37 °C) LysoTracker Green-containing medium.
4. Incubate the cells for 1 h under growth conditions appropriate for the specific cell type.

5. Wash the cells twice with 2 mL of PBS, centrifuge at 350 *g* for 5 min each time, and discard the supernatant. Resuspend the cell pellet with 1 mL of PBS.
6. Add 1 mL of 3 μM DAPI and incubate at room temperature for 15 min.
7. Wash the cells twice with 2 mL of PBS, centrifuge at 350 *g* for 5 min each time, and discard the supernatant.
8. Resuspend the pellet with 400 μL of PBS and analyze it using flow cytometry. Calculate the MFI of LysoTracker Green signals in the live cell populations, as identified by DAPI.

3.3 Flow Cytometry Assay in Ferroptosis

Ferroptosis, first observed in 2003 and formally defined in 2012, is a type of regulated cell death caused by the accumulation of iron-dependent cytotoxic lipid ROS from polyunsaturated fatty acids (PUFA) [17, 18]. The primary mechanism of ferroptosis involves iron catalyzing lipid peroxidation through both non-enzymatic (Fenton reactions) and enzymatic pathways (lipoxygenases) [19, 20]. Autophagy primarily plays a promoting role in ferroptosis, distinct from its role in other forms of RCD [21–23]. These insights have significant implications for cancer therapy [24].

The hallmark characteristics of ferroptosis include the generation of total ROS, lipid ROS, and changes in metal ion levels, particularly iron [25]. The intracellular ROS probe 2′,7-′-dichlorodihydrofluorescein diacetate (DCFH-DA) and the fluorescent probe C11-BODIPY are commonly used to study ferroptosis. C11-BODIPY displays red fluorescence in its normal state, but when its polyunsaturated butadiene part is oxidized, the fluorescence emission peak is transferred from 590 nm to 510 nm, causing the fluorescence color to change from red to green. Liperfluo, an analog of Spy LHP, is used to detect lipid peroxide (LPO) and emits strong fluorescence in organic solvents such as ethanol when oxidized by lipid peroxide. Phen Green SK (PGSK) is a fluorescent heavy metal indicator that reacts with a variety of metal ions, while FerroOrange is a new fluorescence probe used to image Fe^{2+} in living cells.

3.3.1 DCFH-DA Assay

A modified DCFH-DA assay protocol (based on typical cell culture) from MCE (https://www.medchemexpress.cn/h2dcfda.html) is described as follows:

1. Induce ferroptosis or treat the cells using the desired method.
2. Dilute the 5 μM dye probe in pre-warmed PBS to the desired working concentration. The optimal concentration should be determined experimentally.
3. Incubate the cells in the dark for 30 min at 37 °C.

4. Collect the cells using 0.05% trypsin-EDTA, wash with PBS, and centrifuge at 350 *g* for 5 min.
5. Resuspend the cells in 400 μL of PBS containing 5% FBS and determine the MFI of the loaded cell sample using flow cytometry with an excitation/emission wavelength of 488/525 nm.

3.3.2 C11-BODIPY 581/591 Assay

A modified C11-BODIPY 581/591 assay protocol from GlpBio (https://www.glpbio.cn/c11-bodipy-581-591.html) is described as follows:

1. Induce ferroptosis or treat cells using the desired method.
2. Incubate the cells with C11-BODIPY 581/591 in the culture medium at a final concentration of 1–5 μM in the dark for 30 min at 37 °C.
3. Remove excess C11-BODIPY 581/591 by washing the cells twice with PBS.
4. Trypsinize the cells with 0.05% trypsin-EDTA for 1–2 min.
5. Wash the cell pellet twice with PBS, then centrifuge at 350 *g* for 5 min and discard the supernatant.
6. Resuspend the cell pellet in 400 μL of PBS containing 5% FBS.
7. Measure the MFI of the cell sample using flow cytometry. The proportion of MFI (510 nm) to MFI (590 nm) can then be calculated.

3.3.3 Liperfluo Assay

The modified Liperfluo assay protocol (based on typical cell culture) from Dojindo ((https://dojindo.cn/#/goodsDetails?gid=463) is described as follows:

1. Induce ferroptosis or treat cells using the desired method.
2. Harvest cells by adding 0.05% trypsin-EDTA for 1–2 min and transfer them into a 1.5 mL EP tube.
3. Centrifuge the cells at 350 *g* for 5 min and discard the supernatant.
4. Wash the cells with PBS, centrifuge again at 350 *g* for 5 min, and discard the supernatant. Repeat this step once.
5. Add 2 mL of culture medium containing 1–20 μM Liperfluo working solution and incubate in the dark for 30 min at 37 °C.
6. Wash the cells twice with 1 mL PBS, centrifuge at 350 g for 5 min, and discard the supernatant.
7. Resuspend the cell pellet with 400 μL of PBS and analyze the cells on a flow cytometer (Ex = 488 nm, Em = 500–550 nm).

3.3.4 PGSK Assay

A modified Phen Green SK (PGSK) assay protocol, as provided by GlpBio (https://www.glpbio.cn/phen-green-sk-diacetate.html), is described as follows:

1. Induce ferroptosis or treat cells using the desired method.
2. Incubate the cells in a culture medium containing 10 μM Phen Green SK for 10 min in the dark at room temperature.
3. Wash the cells twice with PBS to remove any excess PGSK.
4. Harvest the cells using 0.05% trypsin-EDTA for 1–2 min.
5. Wash the cells twice with 1 mL PBS, centrifuge at 350 *g* for 5 min, and discard the supernatant.
6. Resuspend the cell pellets in 400 μL of PBS and analyze using flow cytometry with an excitation wavelength of 507 nm and emission wavelength of 532 nm.

3.3.5 FerroOrange Assay

A modified FerroOrange assay protocol (based on typical cell culture) from Dojindo (https://dojindo.cn/#/goodsDetails?gid=626) is described as follows:

1. Induce ferroptosis or treat cells by desired method.
2. Harvest the cells by adding 0.05% trypsin-EDTA for 1–2 min and transfer them into a 1.5 mL EP tube.
3. Centrifuge at 350 *g* for 5 min and discard the supernatant.
4. Wash the cells with PBS, centrifuge at 350 *g* for 5 min, and discard the supernatant. Repeat this step once.
5. Add 2 mL of culture medium containing 1–20 μM FerroOrange solution to the cells and incubate in the dark at 37 °C for 30 min.
6. Wash the cells twice with 1 mL PBS, centrifuge at 350 *g* for 5 min, and discard the supernatant.
7. Resuspend the cell pellets in 400 μL of PBS and analyze using flow cytometry (excitation wavelength: 561 nm; emission wavelength: 570–620 nm).

3.4 Flow Cytometry Assay in Pyroptosis

Pyroptosis is a type of regulated cell death that is characterized by the formation of plasma membrane pores by the gasdermin protein family, triggered by the activation of inflammatory caspases. This process leads to the rupturing of the cell membrane and the release of damage-associated molecular patterns (DAMPs), and plays a critical role in the immune response against intracellular pathogens [26–30].

The critical step in pyroptosis is the cleavage of gasdermin by inflammatory caspases, which results in the oligomerization and membrane translocation of the N-terminal fragment of gasdermin [31, 32]. Currently, it is not possible to directly observe these events using flow cytometry. The only way to confirm pyroptosis is by detecting the cleavage of gasdermin D (GSDMD) protein. However, pyroptosis can be indirectly detected by flow cytometry through monitoring the dynamic changes of caspase-1 in a FLICA-caspase 1 assay.

A modified FLICA-caspase 1 assay protocol (based on typical cell culture) from ImmunoChemistry (https://www.immunochemistry.com/products/fam-flica-caspase-1-yvad-assay-kit-97) is described in detail as follows:

1. Induce pyroptosis or treat cells with the desired methods.
2. Reconstruct the lyophilized FLICA vial from the Pyroptosis/Caspase-1 assay kit with 50 μL of DMSO to create a stock solution at a concentration of 150–300×.
3. Dilute the FLICA storage solution with sterile PBS to a working solution concentration of 30–60× by mixing 1 part FLICA solution with 5 parts PBS. Add the working solution to cells in two wells for each condition, with a final concentration of 1–2×.
4. Incubate the cells in the dark at 37 °C for 48 h.
5. Dilute the 10× cell washing buffer from the Pyroptosis/Caspase-1 assay kit to 1× with ddH$_2$O.
6. Transfer the culture medium and dissociated cells to a 5 mL test tube on ice.
7. Add 300 μL of 1× cell washing buffer to the cells and incubate at 37 °C for 10 min. Remove the washing buffer and repeat the transfer to a 5 mL tube.
8. Repeat **step** 7.
9. Add 300 μL of Accutase and incubate at 37 °C for 10–15 min to dissociate all remaining adherent cells. Transfer all the liquid, including dissociated cells, to another 5 mL tube.
10. Wash the cells with 300 μL of 1× cell washing buffer and transfer all the liquid to a 5 mL test tube.
11. Centrifuge at 400 g for 5 min at 4 °C, discard the supernatant, and wash the cells twice with 1× cell washing buffer.
12. Centrifuge the cells again and resuspend them in 300 μL of PBS.
13. Analyze the cells using a flow cytometer and calculate the proportion of FAM-FLICA positive cells (Ex/Em = 488/530 nm).

3.5 Flow Cytometry Assay in Immunogenic Cell Death

Under stress, such as receiving chemotherapy or radiotherapy, tumor cells often undergo both apoptosis and autophagy. This transformation from non-immunogenic to immunogenic cells results in the activation of an anti-tumor immune response, known as immunogenic cell death (ICD) [33, 34]. The latest definition of the Nomenclature Committee on Cell Death states that ICD is a form of RCD that is sufficient to trigger an adaptive immune response in immunocompetent hosts, distinguishing RCDs that drive antigen-specific immune response and result in immune memory from RCDs [1].

DAMPs can be induced and released after ICD. Various factors during immune cell death can cause different DAMP release [35], one of which is calreticulin (CRT). Flow cytometry can detect CRT. CRT is a soluble protein in the endoplasmic reticulum and mediates the regulation of chaperone activity, calcium homeostasis, calcium-mediated signal transduction, correct assembly of MHC class I molecules, and antigen loading [36]. The exposure of CRT to the cell surface serves as an important "eat me" signal, promoting the maturation of naïve dendritic cells and binding to the CD91 receptor in phagocytes to facilitate phagocytosis of dead cells and enhance anti-tumor immune responses [37, 38].

A detailed description of a modified Calreticulin assay protocol (based on typical cell culture) from Cell Signaling Technology (https://www.cellsignal.cn/products/antibody-conjugates/calreticulin-d3e6-xp-rabbit-mab-pe-conjugate/19780?site-search-type=Products&N=4294956287&Ntt=19780&fromPage=plp&_requestid=1430065) is as follows:

1. Treat the cells using the desired methods.
2. Harvest the cells using 0.05% trypsin-EDTA and centrifuge at 350 *g* for 5 min.
3. Discard the supernatant and wash the cells with 1× PBS twice.
4. Resuspend the cells with 4% paraformaldehyde (100 μL/106 cells) and fix the cells at room temperature for 20 min.
5. Centrifuge at 350 *g* for 5 min, discard the supernatant, and wash with 1× PBS once.
6. Discard the supernatant and resuspend the cell pellet with 500–1000 μL of PBS.
7. Add cold 100% methanol while gently stirring the cell suspension until the final concentration of methanol reaches 90%. Permeabilize the cells for at least 10 min on ice.
8. Wash the cells with PBS and centrifuge at 350 *g* for 5 min.
9. Discard the supernatant and resuspend the cell pellet with 100 μL of PBS.
10. Add the fluorescein-conjugated calreticulin mAb to the cell suspension according to the optimized concentration.
11. Incubate the cells in the dark for 60 min at room temperature.
12. Wash with PBS and centrifuge at 350 *g* for 5 min.

13. Discard the supernatant, and resuspend the cell pellet with 400 μL of PBS.
14. Analyze using a flow cytometer (Ex/Em differs according to different conjugated fluorescein).

4 Notes

To ensure successful flow cytometry experiments that yield clean, easy-to-analyze data, extensive planning is required. Each experiment involves many steps, each of which depends on the success of the previous steps. While some steps are common to flow cytometry, others are unique to the experiment or cell model being studied. It's important to keep the following basic principles in mind:

1. Wear appropriate personal protective equipment, such as lab coats, gloves, and safety goggles.
2. Ensure that the equipment is properly maintained and calibrated. Check the laser power and instrument performance before each use.
3. Follow proper sample preparation procedures to avoid contamination or loss of cell viability. Be careful when handling samples to avoid any unnecessary damage to cells.
4. Use appropriate controls, such as unstained cells or isotype controls, to establish baseline measurements and ensure accurate analysis.
5. Dispose of waste properly, including sharps and biohazardous materials. Follow proper disposal protocols for hazardous waste, such as used needles, syringes, and pipette tips.
6. Handle fluorescent dyes and antibodies with care. Store them properly according to manufacturer instructions, and avoid exposure to light and extreme temperatures.
7. Keep the workspace clean and organized to avoid accidental spills or contamination. Clean up spills promptly with appropriate disinfectants.
8. Keep a record of all experimental conditions, including any deviations or modifications from the standard protocol.

References

1. Galluzzi L, Vitale I, Aaronson SA et al (2018) Molecular mechanisms of cell death: recommendations of the Nomenclature Committee on Cell Death 2018. Cell Death Differ 25: 486–541
2. Tang D, Kang R, Berghe TV et al (2019) The molecular machinery of regulated cell death. Cell Res 29:347–364
3. Klionsky DJ, Abdel-Aziz AK, Abdelfatah S et al (2021) Guidelines for the use and interpretation of assays for monitoring autophagy (4th edition). Autophagy:1–382
4. McKinnon KM (2018) Flow cytometry: an overview. Curr Protoc Immunol 120: 5.1–5.1.11

5. Jahan-Tigh RR, Ryan C, Obermoser G et al (2012) Flow cytometry. J Invest Dermatol 132:1–6
6. Manohar SM, Shah P, Nair A (2021) Flow cytometry: principles, applications and recent advances. Bioanalysis 13:181–198
7. Maciorowski Z, Chattopadhyay PK, Jain P (2017) Basic multicolor flow cytometry. Curr Protoc Immunol 117:5.4.1–5.4.38
8. Bertheloot D, Latz E, Franklin BS (2021) Necroptosis, pyroptosis and apoptosis: an intricate game of cell death. Cell Mol Immunol 18: 1106–1121
9. Del Re DP, Amgalan D, Linkermann A et al (2019) Fundamental mechanisms of regulated cell death and implications for heart disease. Physiol Rev 99:1765–1817
10. Su LJ, Zhang JH, Gomez H et al (2019) Reactive oxygen species-induced lipid peroxidation in apoptosis, autophagy, and ferroptosis. Oxidative Med Cell Longev 2019:5080843
11. De Duve C, Wattiaux R (1966) Functions of lysosomes. Annu Rev Physiol 28:435–492
12. Mortimore GE, Schworer CM (1977) Induction of autophagy by amino-acid deprivation in perfused rat liver. Nature 270:174–176
13. Mizushima N, Komatsu M (2011) Autophagy: renovation of cells and tissues. Cell 147:728–741
14. Levine B, Kroemer G (2019) Biological functions of autophagy genes: a disease perspective. Cell 176:11–42
15. Dikic I, Elazar Z (2018) Mechanism and medical implications of mammalian autophagy. Nat Rev Mol Cell Biol 19:349–364
16. Thome MP, Filippi-Chiela EC, Villodre ES et al (2016) Ratiometric analysis of Acridine Orange staining in the study of acidic organelles and autophagy. J Cell Sci 129:4622–4632
17. Tang D, Kroemer G (2020) Ferroptosis. Curr Biol 30:R1–R6
18. Stockwell BR, Friedmann Angeli JP, Bayir H et al (2017) Ferroptosis: a regulated cell death nexus linking metabolism, redox biology, and disease. Cell 171:273–285
19. Chen X, Li J, Kang R et al (2021) Ferroptosis: machinery and regulation. Autophagy 17: 2054–2081
20. Liu J, Kang R, Tang D (2021) Signaling pathways and defense mechanisms of ferroptosis. FEBS J, vol 289, p 7038
21. Hou W, Xie Y, Song X et al (2016) Autophagy promotes ferroptosis by degradation of ferritin. Autophagy 12:1425–1428
22. Li J, Liu J, Xu Y et al (2021) Tumor heterogeneity in autophagy-dependent ferroptosis. Autophagy 17:3361–3374
23. Chen X, Song X, Li J et al (2022) Identification of HPCAL1 as a specific autophagy receptor involved in ferroptosis. Autophagy:1–21
24. Chen X, Kang R, Kroemer G et al (2021) Broadening horizons: the role of ferroptosis in cancer. Nat Rev Clin Oncol 18:280–296
25. Chen X, Comish PB, Tang D et al (2021) Characteristics and biomarkers of ferroptosis. Front Cell Dev Biol 9:637162
26. Yu P, Zhang X, Liu N et al (2021) Pyroptosis: mechanisms and diseases. Signal Transduct Target Ther 6:128
27. Shi J, Gao W, Shao F (2017) Pyroptosis: gasdermin-mediated programmed necrotic cell death. Trends Biochem Sci 42:245–254
28. Kang R, Zeng L, Zhu S et al (2018) Lipid peroxidation drives gasdermin D-mediated pyroptosis in lethal polymicrobial sepsis. Cell Host Microbe 24:97–108. e4, 97
29. Zhang H, Zeng L, Xie M et al (2020) TMEM173 Drives Lethal Coagulation in Sepsis. Cell Host Microbe 27:556–570.e6
30. Tang D, Wang H, Billiar TR et al (2021) Emerging mechanisms of immunocoagulation in sepsis and septic shock. Trends Immunol 42: 508–522
31. Kovacs SB, Miao EA (2017) Gasdermins: effectors of pyroptosis. Trends Cell Biol 27:673–684
32. Shi J, Zhao Y, Wang K et al (2015) Cleavage of GSDMD by inflammatory caspases determines pyroptotic cell death. Nature 526:660–665
33. Galluzzi L, Buque A, Kepp O et al (2017) Immunogenic cell death in cancer and infectious disease. Nat Rev Immunol 17:97–111
34. Galluzzi L, Vitale I, Warren S et al (2020) Consensus guidelines for the definition, detection and interpretation of immunogenic cell death. J Immunother Cancer 8
35. Fucikova J, Kepp O, Kasikova L et al (2020) Detection of immunogenic cell death and its relevance for cancer therapy. Cell Death Dis 11: 1013
36. Fucikova J, Spisek R, Kroemer G et al (2021) Calreticulin and cancer. Cell Res 31:5–16
37. Li W, Yang J, Luo L et al (2019) Targeting photodynamic and photothermal therapy to the endoplasmic reticulum enhances immunogenic cancer cell death. Nat Commun 10:3349
38. Obeid M, Panaretakis T, Tesniere A et al (2007) Leveraging the immune system during chemotherapy: moving calreticulin to the cell surface converts apoptotic death from "silent" to immunogenic. Cancer Res 67:7941–7944

Chapter 16

Thermal Shift Assay in Ferroptosis

Sridhar Bammidi, Stacey Hose, James T. Handa, Debasish Sinha, and Sayan Ghosh

Abstract

Ferroptosis is a recently described process of cell death that is dependent on unregulated cellular iron accumulation with induction of oxidative stress. Ferroptosis has been linked to several human diseases; therefore, investigations aimed at better understanding the pathway and elucidating avenues for future drug development are warranted. Current assays that target ferroptosis/oxidative stress in cells is limited to western blotting and imaging techniques, and unfortunately provide only a broad understanding that is insufficient to effectively assess novel drugs (ligands). Specifically, these assays do not provide insights about ligand interactions with specific proteins associated with these processes. Herein, we discuss a cell-based thermal shift assay that enables screening of ligands under specific cellular conditions for targeting ferroptosis and/or oxidative stress pathways. These data would provide detailed preliminary evidence required for drug development that targets this pathway.

Key words Ferroptosis, Cells, Thermal shift assay, Oxidative stress, Iron

1 Introduction

Ferroptosis, iron-induced cell death, has recently been identified as a mode of cell death that is induced by the unregulated accumulation of cellular iron [1]. Increased iron accumulation triggers redox imbalance that induces lipid peroxidation, and ferroptosis [1]. Several lipid peroxidation intermediates and products are increased during ferroptosis like hydroperoxides (LOOHs) and reactive aldehydes such as malondialdehyde (MDA) and 4-hydroxynonenal (4HNE) [1, 2]. Fatty acid oxidation is considered the major trigger of ferroptosis [1]. Inadequately neutralized oxidative stress modifies fatty acids to alter cellular membrane integrity and provides the messengers for ferroptosis induction [3]. Specifically, cell membrane lipids like phosphatidylcholine, phosphatidylethanolamine (PE), and cardiolipin are one of the major sources of MDA formation due to elevated cellular levels of reactive oxygen species and reactive nitrogen species [1]. In addition, peroxidation of

Guido Kroemer and Daolin Tang (eds.), *Ferroptosis: Methods and Protocols*, Methods in Molecular Biology, vol. 2712,
https://doi.org/10.1007/978-1-0716-3433-2_16,

polyunsaturated fatty acids (PUFAs) in phospholipids by arachidonic acid lipoxygenases (ALOXs) appears to be of particular importance for ferroptosis [4, 5].

Several signaling molecules are associated with ferroptosis [1]. In cells, iron levels are controlled by ferritin, which contains a complex of Ferritin Heavy Chain 1 (FTH1) and Light (FTL) chains [6]. Also, level of ferritin in cells is regulated by selective autophagy that targets ferritin, termed ferritinophagy, which is in turn regulated by nuclear receptor coactivator 4 (NCOA4), a selective cargo receptor for ferritin [7, 8]. The divalent metal transporter 1 (DMT1) also regulates iron homeostasis through non-heme absorption [9, 10]. In addition, the glutathione peroxidase pathway has been identified as a key redox-regulating pathway involved in ferroptosis. RAS-selective lethal 3 (RSL3), which is a specific inhibitor of glutathione peroxidase 4 (GPX4), can activate ferroptosis [11]. Glutathione (GSH) is converted into oxidized glutathione (GSSH) by GPX4 and reduces cytotoxic lipid peroxides [12]. The GPX4 pathway is also regulated by System Xc-, an amino acid antiporter consisting of a disulfide-linked heterodimer of SLC7A11/xCT and SLC3A2/4F2hc/CD98, which is inhibited by the ferroptosis inducer erastin [13]. The regulation of genes involved in oxidative stress, including GPX4, is largely controlled by the transcription factor nuclear factor erythroid 2–related factor 2 (NRF2), which serves as a defense against ferroptosis [14]. Under normal conditions, NRF2 is sequestered in the cytosol through its interaction with Kelch-like ECH-associated protein 1 (KEAP1), which is part of a ubiquitin E3 ligase complex that leads to NRF2 proteasomal degradation [15]. Oxidative stress leads to conformational changes in KEAP1 to thereby disrupt its interaction with NRF2, which migrates into the nucleus and activates NRF2-mediated antioxidant genes [15]. This process is further regulated through the autophagy pathway whereby the autophagy cargo receptor p62/SQSTM1 can initiate competitive inhibition on the KEAP1-NRF2 complex, leading to upregulation of NRF2 [16, 17].

Recent studies have shown that ferroptosis is closely related to the pathophysiological processes of many diseases, such as tumors [18], nervous system diseases [19, 20], ischemia-reperfusion injury [21, 22], kidney injury [23, 24], retinal degeneration [25, 26], pancreatitis [27], and blood diseases [28]. Therefore, identifying molecules that can target the ferroptosis process may be a viable approach to preventing or inhibiting the progression of these diseases.

To date, assays to determine the state of ferroptosis activation or inhibition in cells under different conditions have been documented either by western blotting for ferroptosis markers like FTH1, GPX4, NRF2, and DMT1, among others [10, 14, 26], or by imaging studies using liperfluo dye, which is a perylene derivative containing oligooxyethylene, that detects lipid peroxides [26]. Unfortunately, these assays lack the specificity and sensitivity needed for initial drug screening that provide knowledge about the binding of ligands to particular protein targets in the ferroptosis/ oxidative stress pathway. Moreover, established assays that screen drugs for targeting proteins involved in ferroptosis and/or oxidative stress are limited.

Herein, we will discuss the utilization of cellular thermal shift assay (CETSA) to identify putative molecules that can target/bind proteins involved with the induction of ferroptosis and oxidative stress pathways (*see* **Note 1**). It is a method to screen ligands for target protein binding within a particular cellular environment by detecting changes in a protein's thermal stability upon ligand binding [29]. The method utilizes basic heat-denaturation followed by western blotting to ascertain the binding affinity of the ligand with the target protein of interest (*see* **Note 2**).

2 Materials

1. Cells of interest (any adherent/suspended cells or explants can be used).
2. Ligand/molecule to test (purified).
3. CO_2 incubator.
4. Biosafety Cabinet.
5. MicroAmp Fast Reaction Tubes: 8 tubes/strip (Applied Biosystems, USA).
6. Eppendorf Safe-Lock Microcentrifuge Tubes, Polypropylene, Tube (Fisher Scientific, USA).
7. RIPA 10× lysis buffer (EMD Millipore, USA).
8. Protease Inhibitor Cocktail Animal Component Free (Sigma Aldrich, USA).
9. Phosphatase Inhibitor Cocktail 3, DMSO solution (Sigma Aldrich, USA).
10. Pierce BCA Protein Assay Kit (Thermo Fisher Scientific, USA).
11. Thermal cycle-Veriti (Applied Biosystems, USA): Can be changed to individual need.
12. Nu-PAGE 4–12% Bis-TRIS Gel-15 well (Invitrogen, USA): Can be changed to individual need.
13. iBLOT2 NC Regular Stacks (Invitrogen, USA): Can be changed to individual need.
14. iBlot2 transfer apparatus (Invitrogen, USA): Can be changed to individual need.
15. Gel Apparatus (Invitrogen, USA): Can be changed to individual need.

3 Methods

The method is depicted in a stepwise manner (Fig. 1).

1. Cells are grown to 85% confluency and can be used to either induce ferroptosis with RSL3 or erastin, or left untreated (*see* **Note 3**).
2. The cells are lysed using 1X RIPA containing 1% each of phosphatase and protease inhibitor cocktails.
3. The protein levels in the lysates are then quantified using a BCA kit and 20 mg of protein is aliquoted into PCR tubes and treated with the ligand of choice (vehicle is used as control) for 1 h at room temperature prior to CETSA heat pulse (*see* **Note 4**).

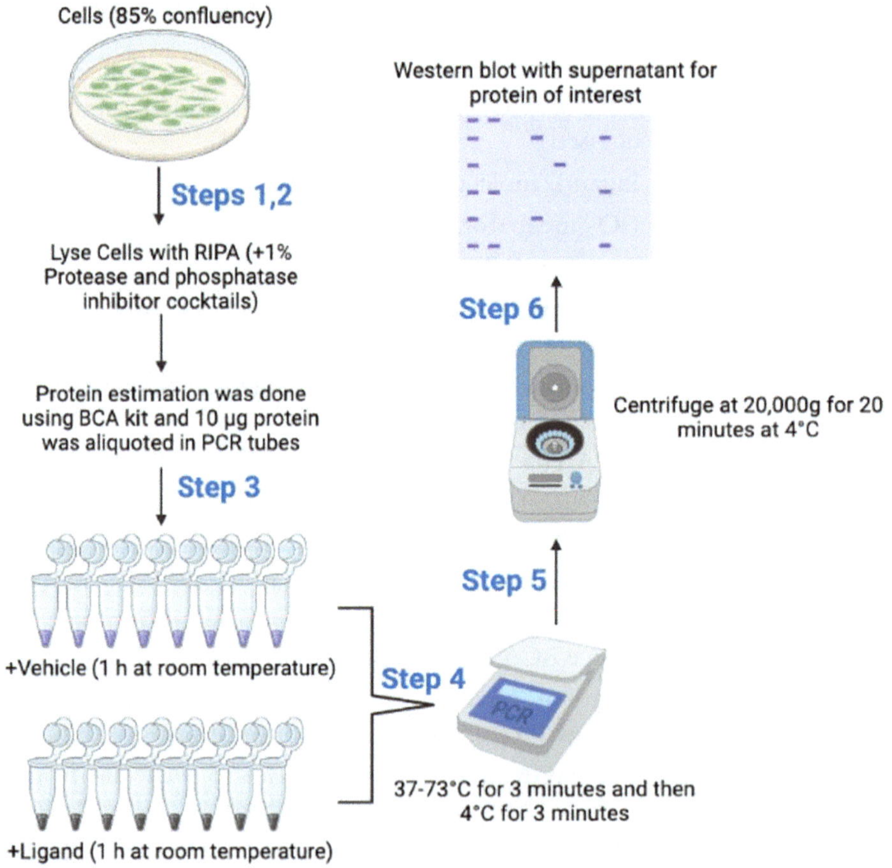

Fig. 1 Method (step-wise) of cellular thermal shift assay (CETSA). The illustration was created by BioRender.com accessed on April 1, 2023

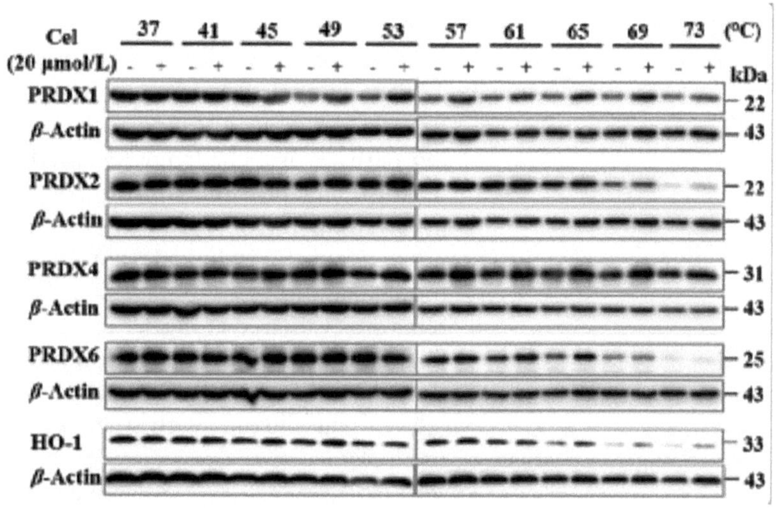

Fig. 2 Fig. 4D used with permission from Acta Pharm Sin B. 12(5). Luo P, Liu D, Zhang Q, Yang F, Wong YK, Xia F, Zhang J, Chen J, Tian Y, Yang C, Dai L, Shen HM, Wang J. Celastrol induces ferroptosis in activated HSCs to ameliorate hepatic fibrosis via targeting peroxiredoxins and HO-1. 2300–2314. Copyright Elsevier (2022) [31]
Western blot showing expression of several proteins with celastrol (20 mmol/L) or DMSO treatment to LX-2 cells with increasing temperature. Proteins like PRDX1, PRDX2, PRDX4, PRDX6, and HO-1 all displayed significant thermal stabilization (higher expression, compared to control at high temperature) in the celastrol treatment group

4. The solutions are heated at the indicated temperatures of 37, 41, 45, 49, 53, 57, 61, 65, 69, and 73 °C for 3 min each, followed by cooling at 4 °C for 3 min in a thermal cycler (Applied Biosystems, USA) (*see* **Note 4**).
5. The tubes are then centrifuged for 20 min (20,000× *g*, 4 °C), the soluble supernatant is subjected to western blotting [26, 30] (*see* **Note 4**). Target protein abundance estimation is performed using the specific primary antibody, as shown in the example in Fig. 2 (used with permission from *Acta Pharm Sin B. 12(5). Luo P, Liu D, Zhang Q, Yang F, Wong YK, Xia F, Zhang J, Chen J, Tian Y, Yang C, Dai L, Shen HM, Wang J. Celastrol induces ferroptosis in activated HSCs to ameliorate hepatic fibrosis* via *targeting peroxiredoxins and HO-1. 2300–2314. Copyright Elsevier (2022)*) [31].
6. The western blot data are quantified using Image J software [26, 30] to ascertain the thermal stability of the protein of interest in the presence or absence of the ligand (as shown in Fig. 2) (*see* **Note 4**).

4 Notes

1. To date, very little is known about ferroptosis in disease pathogenesis [1]. Thus, drug development for such an elusive signaling pathway is a challenge [32]. CETSA can be used to screen multiple ligands of interest for targeting specific proteins involved in ferroptosis with a cell-based approach, which is more physiologically relevant than cell-free thermal shift or molecular biology-based assays. Briefly, levels of specific proteins are evaluated at higher temperatures in the presence of the ligand or vehicle (control). A higher amount of protein (more thermal stability) in ligand-treated tubes indicates strong binding of the ligand with the protein, compared to vehicle controls.
2. The method identifies the ability of a particular ligand that can bind to mediators of ferroptosis and oxidative stress in cells. However, simultaneous experiments (western blot or imaging studies) must be performed to demonstrate whether binding of the ligand regulates the levels of the target protein and ferroptosis pathway activity.
3. Since this assay is cell-based, several conditions can be tested as needed (Methods; **step 1**), therefore, this assay provides a great deal of flexibility in screening ligands to target proteins under different cellular conditions. This assay also has the potential to be used to screen drugs for a multitude of cellular processes other than ferroptosis and oxidative stress.
4. Finally, this assay protocol can be modified. If the ligand can penetrate cells and is stable over a long period of time, cells can be pre-treated with the ligand of choice, and then the above steps are performed (Methods; **steps 3–6**).

Acknowledgments

This work is supported by NIH R01EY031594 (DS and JTH), R01EY032516 (DS) and BrightFocus Postdoctoral Fellowship on Macular Degeneration (SG), P30 core award EY08098 from the National Eye Institute, NIH, and unrestricted funds from The Research to Prevent Blindness Inc., NY, and the Eye & Ear Foundation of Pittsburgh (to the University of Pittsburgh Department of Ophthalmology).

References

1. Tang D, Chen X, Kang R, Kroemer G (2021) Ferroptosis: molecular mechanisms and health implications. Cell Res 31(2):107–125
2. Chen X, Huang J, Yu C, et al (2022). A noncanonical function of EIF4E limits ALDH1B1 activity and increases susceptibility to ferroptosis. Nat Commun13(1):6318
3. Tang D, Kroemer G, Kang R (2023) Ferroptosis in immunostimulation and immunosuppression. Immunol Rev doi: 10.1111/imr.13235. Epub ahead of print.
4. Kagan VE, Mao G, Qu F, Angeli JP, Doll S, Croix CS et al (2017) Oxidized arachidonic and adrenic PEs navigate cells to ferroptosis. Nat Chem Biol 13(1):81–90
5. Wenzel SE, Tyurina YY, Zhao J, St Croix CM, Dar HH, Mao G et al (2017) PEBP1 wardens ferroptosis by enabling lipoxygenase generation of lipid death signals. Cell 171(3):628–641.e26
6. Chen X, Yu C, Kang R, Tang D (2020) Iron metabolism in Ferroptosis. Front Cell Dev Biol 8:590226
7. Mancias JD, Wang X, Gygi SP, Harper JW, Kimmelman AC (2014) Quantitative proteomics identifies NCOA4 as the cargo receptor mediating ferritinophagy. Nature 509(7498):105–109
8. Dowdle WE, Nyfeler B, Nagel J, Elling RA, Liu S, Triantafellow E et al (2014) Selective VPS34 inhibitor blocks autophagy and uncovers a role for NCOA4 in ferritin degradation and iron homeostasis in vivo. Nat Cell Biol 16(11):1069–1079
9. Gunshin H, Mackenzie B, Berger UV, Gunshin Y, Romero MF, Boron WF et al (1997) Cloning and characterization of a mammalian proton-coupled metal-ion transporter. Nature 388(6641):482–488
10. Song Q, Peng S, Sun Z, Heng X, Zhu X (2021) Temozolomide drives Ferroptosis via a DMT1-dependent pathway in glioblastoma cells. Yonsei Med J 62(9):843–849
11. Yang WS, SriRamaratnam R, Welsch ME, Shimada K, Skouta R, Viswanathan VS et al (2014) Regulation of ferroptotic cancer cell death by GPX4. Cell 156(1–2):317–331
12. Li J, Cao F, Yin HL, Huang ZJ, Lin ZT, Mao N et al (2020) Ferroptosis: past, present and future. Cell Death Dis 11(2):88
13. Dixon SJ, Patel DN, Welsch M, Skouta R, Lee ED, Hayano M et al (2014) Pharmacological inhibition of cystine-glutamate exchange induces endoplasmic reticulum stress and ferroptosis. elife 3:e02523
14. Sun X, Ou Z, Chen R, et al (2016) Activation of the p62-Keap1-NRF2 pathway protects against ferroptosis in hepatocellular carcinoma cells. Hepatology 63(1):173–84
15. Baird L, Yamamoto M (2020) The molecular mechanisms regulating the KEAP1-NRF2 pathway. Mol Cell Biol 40(13):e00099–e00020
16. Lee DH, Park JS, Lee YS, Han J, Lee DK, Kwon SW et al (2020) SQSTM1/p62 activates NFE2L2/NRF2 via ULK1-mediated autophagic KEAP1 degradation and protects mouse liver from lipotoxicity. Autophagy 16(11):1949–1973
17. Sun X, Ou Z, Chen R, Niu X, Chen D, Kang R et al (2016) Activation of the p62-Keap1-NRF2 pathway protects against ferroptosis in hepatocellular carcinoma cells. Hepatology 63(1):173–184
18. Tang D, Kroemer G, Kang R (2023) Ferroptosis in hepatocellular carcinoma: from bench to bedside. Hepatology. doi: 10.1097/HEP.0000000000000390. Epub ahead of print.
19. Zhou Y, Lin W, Rao T, Zheng J, Zhang T, Zhang M et al (2022) Ferroptosis and its potential role in the nervous system diseases. J Inflamm Res 15:1555–1574
20. Lan T, Sun TT, Wei C, Cheng T, Yang F, Zhang JN et al (2023) Epigenetic regulation of ferroptosis in central nervous system diseases. Mol Neurobiol 60:3584
21. Zhao WK, Zhou Y, Xu TT, Wu Q (2021) Ferroptosis: opportunities and challenges in myocardial ischemia-reperfusion injury. Oxidative Med Cell Longev 2021:9929687
22. Li X, Ma N, Xu J, Zhang Y, Yang P, Su X et al (2021) Targeting Ferroptosis: pathological mechanism and treatment of ischemia-reperfusion injury. Oxidative Med Cell Longev 2021:1587922
23. Wang Y, Quan F, Cao Q, Lin Y, Yue C, Bi R et al (2020) Quercetin alleviates acute kidney injury by inhibiting ferroptosis. J Adv Res 28:231–243
24. Wang Y, Zhang M, Bi R, Su Y, Quan F, Lin Y et al (2022) ACSL4 deficiency confers protection against ferroptosis-mediated acute kidney injury. Redox Biol 51:102262
25. Tang W, Guo J, Liu W, Ma J, Xu G (2021) Ferrostatin-1 attenuates ferroptosis and protects the retina against light-induced retinal degeneration. Biochem Biophys Res Commun 548:27–34

26. Gupta U, Ghosh S, Wallace CT, Shang P, Xin Y, Nair AP et al (2023) Increased LCN2 (lipocalin 2) in the RPE decreases autophagy and activates inflammasome-ferroptosis processes in a mouse model of dry AMD. Autophagy 19(1):92–111
27. Wei S, Qiu T, Yao X, Wang N, Jiang L, Jia X et al (2020) Arsenic induces pancreatic dysfunction and ferroptosis via mitochondrial ROS-autophagy-lysosomal pathway. J Hazard Mater 384:121390
28. Chen Z, Jiang J, Fu N, Chen L (2022) Targeting ferroptosis for blood cell-related diseases. J Drug Target 30(3):244–258
29. Owens AE, Iannotti MJ, Sanchez TW, Voss T, Kapoor A, Hall MD et al (2022) High-throughput cellular thermal shift assay using acoustic transfer of protein lysates. ACS Chem Biol 17(2):322–330
30. Ghosh S, Liu H, Yazdankhah M, Stepicheva N, Shang P, Vaidya T et al (2021) βA1-crystallin regulates glucose metabolism and mitochondrial function in mouse retinal astrocytes by modulating PTP1B activity. Commun Biol 4(1):248
31. Luo P, Liu D, Zhang Q, Yang F, Wong YK, Xia F et al (2022) Celastrol induces ferroptosis in activated HSCs to ameliorate hepatic fibrosis via targeting peroxiredoxins and HO-1. Acta Pharm Sin B 12(5):2300–2314
32. Hadian K, Stockwell BR (2021) A roadmap to creating ferroptosis-based medicines. Nat Chem Biol 17(11):1113–1116

Chapter 17

Detection of Ferroptosis in Patient-Derived Tumor Models

Wenjing Zhang and Yang Wang

Abstract

Ferroptosis is a form of iron-dependent regulated cell death that is primarily caused by the accumulation of iron, lipid peroxidation, and subsequent rupture of the plasma membrane. The process and function of ferroptosis can be monitored in multiple ways, both in vitro and in vivo. Patient-derived xenograft (PDX) is a type of preclinical cancer model that involves transplanting human cancer tissue, usually obtained from patients undergoing surgery or biopsy, into immunodeficient mice or other animal models. It is a powerful tool for understanding drug response in cancer, as PDX models preserve the growth environment and heterogeneity of the original tumors. By analyzing ferroptosis in PDX models, we can potentially gain insights into human tumorigenesis. In this article, we summarize several assays used to analyze ferroptosis in PDX models.

Key words Ferroptosis, Patient-derived xenograft, Transmission electron microscope, Immunohistochemistry, Iron assay, Western blotting

1 Introduction

Ferroptosis is a form of iron-dependent cell death that differs from other types of regulated cell death, including apoptosis, pyroptosis, and necroptosis [1]. It is primarily driven by iron accumulation, lipid peroxidation, and subsequent plasma membrane rupture [2]. Several critical factors modulate ferroptosis by regulating the oxidation state of cellular lipids and iron content, including glutathione peroxidase 4 (GPX4), solute carrier family 7 member 11 (SLC7A11), acyl-CoA synthetase long-chain family member 4 (ACSL4), ferritin, transferrin, and transferrin receptor 1 [3]. Among them, GPX4 is the main repressor of ferroptosis that uses glutathione (GSH) to resist lipid peroxidation [4, 5]. SLC7A11 is a key component of system xc^- and transports cystine into cells to generate GSH [1]. Inhibition of SLC7A11 or GPX4 by erastin or RSL3, respectively, is a classic model for inducing ferroptosis [1–4, 6]. ACSL4 converts free long-chain fatty acids into fatty acetyl-CoA esters, thereby participating in lipid

Guido Kroemer and Daolin Tang (eds.), *Ferroptosis: Methods and Protocols*, Methods in Molecular Biology, vol. 2712, https://doi.org/10.1007/978-1-0716-3433-2_17,

biosynthesis and fatty acid degradation [7, 8]. Changes in the protein expression of SLC7A11, GPX4, or ACSL4 can be used as markers to predict cell sensitivity or resistance to ferroptosis [9].

Iron plays a crucial role in the process of ferroptosis [10]. The level of intracellular or mitochondrial ferrous iron is increased in ferroptotic cells or tissues [11]. Iron accumulation can cause lipid peroxidation, leading to subsequent membrane damage and rupture [1]. Iron chelators can effectively prevent ferroptosis in vitro or in vivo [1]. Multiple regulators of iron metabolism are involved in controlling the sensitivity of ferroptosis by regulating iron uptake, storage, utilization, and efflux. Transferrin, lactotransferrin, and transferrin receptor (TFR) mediate iron uptake, which can trigger ferroptosis [12–14]. TFR is considered a biomarker of ferroptosis in cells and tissues [15].

In vitro studies in cells derived from patients may not recapitulate the complexity of tumors because they substantially deviate from the original biopsy upon culturing and do not predict the response of the organism as a whole. Therefore, patient-derived xenograft (PDX) models are established by implanting freshly isolated tumors from patients into immunocompromised mice, allowing the tumor to grow in vivo [16]. PDX models provide powerful tools to understand the role of ferroptosis in cancer because they simulate an in vivo environment, preserve the heterogeneity of the original tumors, and are more reliable in predicting patient response to therapeutics than tissue culture-based methods. Thus, analyzing ferroptosis in PDX models can potentially offer information that can be extrapolated to tumorigenesis in humans.

In this chapter, we introduce several methods to detect ferroptosis in PDX models based on the characteristics of morphology and metabolism of ferroptosis.

2 Materials

1. 10% formalin.
2. 2.5% glutaraldehyde in 0.1 M sodium cacodylate buffer, pH 7.4.
3. 0.1 M PBS buffer.
4. 1% buffered osmium.
5. Acetone and epoxy resin (1:2).
6. 50–90% ethanol.
7. TEM grids 200 mesh gold 3.05 mm.
8. 0.5% uranyl acetate.
9. 3% lead citrate.
10. Xylene.

11. Paraffin Wax.
12. 0.01 M sodium citrate buffer.
13. 3% H_2O_2.
14. Blocking buffer (10% normal goat serum with 0.1% Triton X-100 and 3% BSA in PBS).
15. Hematoxylin.
16. Universal SAP kit (ZSGB-Bio, SAP-9100).
17. DAB color development kit (ZSGB-Bio, ZLI-9018).
18. Iron assay kit (abcam, ab83366).
19. 0.1% Triton X-100.
20. RIPA Lysis Buffer (25 mM Tris-HCl, pH 7.6; 150 mM NaCl; 1% NP-40; 1% sodium deoxycholate; 0.1% SDS).
21. Liquid nitrogen.
22. Oven.
23. Transmission electron microscope.
24. Heat-proof slide container.
25. Microwave.
26. Bright field microscope.
27. Colorimetric microplate reader.
28. Microtome.
29. Mini PROTEAN system.
30. Homogenizer.
31. Antibodies: such as anti-GPX4.
32. Digital imaging systems or X-ray film cassette.

3 Methods

3.1 Transmission Electron Microscopy Assay

The transmission electron microscope (TEM) is an electron microscope that uses a focused beam of electrons to produce highly magnified and detailed images of a specimen. It is a powerful tool for studying the ultrastructure of cells and tissues, and is considered the gold standard for determining the types of cell death.

In particular, TEM can distinguish among different forms of cell death, such as ferroptosis, apoptosis, and necroptosis, based on characteristic changes in cell morphology. For example, in cells undergoing ferroptosis, the mitochondria exhibit smaller size compared to normal mitochondria, with condensed mitochondrial membrane densities, reduction or disappearance of mitochondria cristae, and outer mitochondrial membrane rupture. These features are distinct from those seen in other forms of cell death.

3.1.1 Fixation (see Note 1)

1. Sacrifice the mice that are growing patient-derived tumor tissue. Excise the tumor and trim away the fascia and membranous outer layer surrounding the tumor. Section the tumor into smaller pieces for different uses: fix in TEM fixing buffer for TEM; fix in 10% formalin for immunohistochemistry; use fresh tissue for iron assay; flash-freeze in liquid nitrogen for protein extraction.
2. Fix tumor tissue (~1 mm^3) with 2.5% glutaraldehyde in 0.1 M sodium cacodylate buffer, pH 7.4, at 4 °C overnight.
3. Wash the tissue three times for 15 min with 0.1 M PBS buffer.
4. Postfix the tissue with 1% buffered osmium for 2 h.

3.1.2 Dehydration

1. Wash the tissue three times for 15 min with 0.1 M PBS buffer.
2. Dehydrate the samples by incubating once with 50%, 70%, 80%, and 90% ethanol, and twice with 100% ethanol, each step for 15 min.

3.1.3 Embedding

1. After dehydration, incubate the samples with acetone and epoxy resin (1:2) at 4 °C overnight to allow embedding resin to infiltrate into the tissues.
2. Place the samples in an ultramicrotome holder and embed them into epoxy resin. Incubate in a 37 °C oven overnight.
3. Polymerize the resin in a 45 °C oven for 12 h and then in a 60 °C oven for 48 h.

3.1.4 Sectioning (see Note 2)

1. Using an ultramicrotome and a diamond knife, cut 70 nm ultrathin sections.
2. Collect ultrathin sections on metal-mesh "grids."
3. Dry sections overnight before staining.

3.1.5 Staining (see Note 3)

1. Place the grids floating on a drop of 0.5% uranyl acetate with the section side down and incubate in the dark for 30 min at room temperature. Thoroughly wash the grids by dipping them into a beaker filled with fresh double-distilled water.
2. Place the grids floating on a drop of 3% lead citrate with the section side down and incubate in the dark for 30 s at room temperature. Thoroughly wash the grids by dipping them into a beaker filled with fresh double-distilled water.
3. Dry the grid by touching with filter paper on the edge of the grid.
4. Use a transmission electron microscope to examine cell morphology, especially mitochondria.

3.2 Immuno-histochemistry Assay

Lipid peroxidation is a process in which oxidants attack lipids, and it plays a central role in mediating ferroptosis [6]. The oxidation of polyunsaturated fatty acids (PUFAs) by reactive oxygen species (ROS) to produce lipid hydroperoxides is a key step in driving ferroptosis [17]. Malondialdehyde (MDA) and 4-hydroxynonenal (4HNE) are the final products of lipid peroxidation of PUFAs and serve as biomarkers of ferroptosis [3].

Immunohistochemistry (IHC) can be used to quantify MDA (1:400, AdipoGen, JAI-MMD-030N) and 4HNE (1:200, Abcam, ab46545) adducts to determine the extent of lipid peroxidation of ferroptotic cells in tissue samples. IHC combines anatomical, immunological, and biochemical techniques to image the target protein in tissues by using a specific primary antibody to bind their target antigens in situ.

3.2.1 Tissue Fixation

1. Immediately after isolating fresh tissue, cut a piece of tumor tissue using scissors. Holding the tissue with forceps, immerse the piece in PBS to remove blood, fur, etc.
2. Immediately place the rinsed tissue in a labeled tissue cassette and fix the tissue in 10% formalin overnight.
3. Wash 3 × 10 min with an ample amount of PBS at room temperature.

3.2.2 Tissue Embedding in Paraffin Wax (see **Note 4**)

1. Dehydrate the tissue through graded ethanol (70%, 85%, 95%, 100%, 20 min each) and incubate twice with xylene (30 min each).
2. Infiltrate tissue in molten paraffin wax (approximately 56 °C) three times for a total of 1 h.

3.2.3 Paraffin Wax Sectioning (see **Note 5**)

1. Section the wax sample at 3–10 μm thickness using a microtome. 5 μm is the most commonly used thickness.
2. Place the wax ribbons in a water bath at 45 °C to flatten the sections.
3. Collect the sections on positively charged slides.
4. Dry the slides for 60 min at 60 °C.

3.2.4 Deparaffinization/Rehydration

1. Deparaffinize the tissue sections by incubating slides in two washes of xylene for 5 min each.
2. Hydrate the tissue sections by incubating the slides in two washes of 100% ethanol for 5 min each, 95%, 80%, 70%, and 50% ethanol for 3 min each.
3. Place the slide rack in PBS to complete the hydration process.

*3.2.5 Antigen Retrieval (see **Note 6**)*

1. Place hydrated slides in a heat-proof slide container filled with 0.01 M sodium citrate buffer. Heat slides in a microwave for 5 min.
2. Place the container at room temperature and cool the slides for 20 min.
3. Place the container in a microwave and heat for 3 min.
4. Place the container at room temperature and cool the slides for 15 min.
5. Rinse slides three times with 0.1 M PBS for 5 min each.

*3.2.6 Blocking (see **Note 7**)*

1. Incubate sections in 3% H_2O_2 for 30 min at room temperature to quench endogenous peroxidase.
2. Rinse slides three times in 0.1 M PBS for 5 min each.
3. Lay the slides on the humidified chamber and add 100–400 μL blocking buffer (10% normal goat serum with 0.1% Triton X-100 and 3% BSA in PBS) to cover the slides for 1 h at room temperature.

3.2.7 Immuno-histochemical Staining with Universal SAP Kit (ZSGB-Bio, SAP-9100)

1. Remove blocking buffer and apply 100–400 μL of primary antibody diluted in blocking buffer (at the appropriate dilution recommended in the product datasheet).
2. Incubate the slides overnight at 4 °C in a humidified chamber.
3. Remove the antibody solution and rinse the slides three times with PBS for 5 min each with gentle agitation.
4. Add 100–400 μL of biotinylated goat anti-mouse/rabbit IgG solution to the section and incubate for 1 h at room temperature.
5. Rinse the slides three times with PBS for 5 min each with gentle agitation.
6. Add 100–400 μL of Streptavidin-HRP conjugate detection complex to the section and incubate for 1 h at room temperature.
7. Rinse the slides three times with PBS for 5 min each with gentle agitation.
8. Add 100–400 μL of freshly prepared DAB chromogenic solution using DAB color development kit (ZSGB-Bio, ZLI-9018) to the slides and allow the color to develop for 1–10 min. Use a light microscope to monitor the development and intensity of the staining.
9. Rinse the slides in running tap water for 5 min to stop the reaction.

3.2.8 Nuclear Counterstaining

1. Submerge the slides in hematoxylin and incubate for 1 min.
2. Rinse with tap water until the rinse water is clear.
3. Rinse with PBS for 5 min.
4. Dehydrate the sections through graded ethanol (50%, 70%, 95% for 3 min each, and twice with 100% for 3 min each).
5. Incubate the slides twice with xylene for 3 min each.

3.2.9 Mounting Sections and Sample Visualization

1. Apply mounting medium to cover the section and place a coverslip over it. Take care to avoid air bubbles. Air dry at room temperature.
2. Visualize the tissue staining under a bright field microscope.

3.3 Iron Assay

Iron is a vital trace element that is stored in the center of metalloproteins, in the heme complex, and in oxygen carrier proteins. In mammalian cells, iron exists in two forms: ferrous iron (Fe^{2+}) and ferric iron (Fe^{3+}). As iron is essential for ferroptosis, and the level of ferrous iron is elevated in ferroptotic cells or tissues, the detection of iron content in cells or tissues is a commonly used method for monitoring ferroptosis.

The commercial iron assay kit provides a simple and convenient means of measuring ferrous iron and/or ferric iron in biological samples. Free ferrous iron reacts with Iron Probe to produce a stable-colored complex with an absorbance at 593 nm, thereby determining the level of ferrous iron. Under acidic conditions, ferric iron is dissociated from ferric carrier proteins, released into the solution, and reduced to form Fe^{2+}, enabling the measurement of total iron (Fe^{2+} and Fe^{3+}). The level of ferric iron (Fe^{3+}) is calculated by subtracting the ferrous iron from the total iron measurement.

3.3.1 Sample Preparation (see **Note 8**)

1. Harvest 10–100 mg of tumor tissue from PDX mice.
2. Rinse the tissue with cold PBS.
3. Use a Dounce homogenizer to homogenize the tissue for 10–15 passes with 4–10 volumes of Iron Assay Buffer on ice.
4. Centrifuge at 16,000 *g* for 10 min to remove insoluble materials.
5. Collect the supernatant and transfer it to a clean tube. Keep it on ice.

3.3.2 Standard Preparation

1. Prepare a 1 mM standard by diluting 10 μL of Iron Standard (100 mM) in 990 μL of ddH_2O.
2. Using the 1 mM standard, prepare standard curve dilutions (final volume of standard in well is 100 μL, and end concentrations of iron in well are 0, 2, 4, 6, 8, and 10 nmol/well).

3.3.3 Assay Procedure

1. Add 100 μL of standard dilutions to the standard wells.
2. Add 2–50 μL of samples to the sample wells and adjust the volume to 100 μL/well with Iron Assay Buffer.
3. Add 5 μL of iron reducer to each standard well.
4. Add 5 μL of assay buffer to each sample well to determine ferrous iron (Fe^{2+}).
5. Mix and incubate the standards and samples at 37 °C for 30 min.
6. Add 100 μL of Iron Probe to each well.
7. Mix and incubate at 37 °C for 1 h protected from light.
8. Measure absorbance immediately on a colorimetric microplate reader at 593 nm.
9. Draw the iron standard calibration curve and calculate the content of ferrous iron (Fe^{2+}).

3.4 Western Blotting

Western blotting (WB), also known as immunoblotting, is a widely used and powerful laboratory technique for the immunological detection of proteins. It involves the separation of complex mixtures of proteins based on their charge and molecular weight via gel electrophoresis, followed by their transfer to a polyvinylidene fluoride (PVDF) membrane. The membrane is then incubated with a primary antibody specific to the target protein. After incubation, the unbound antibody is washed away, and the membrane is incubated with a horseradish peroxidase (HRP)-tagged secondary antibody that recognizes the primary antibody. The target protein can be detected using chemiluminescence with traditional film or digital imaging systems. WB can be used to measure the levels of protein markers, such as GPX4 (Proteintech 67763-1-Ig, 1: 1000), SLC7A11 (CST 12691S), and ACSL4 (SANTA SC-271800), to monitor the process of ferroptosis in PDX models.

3.4.1 Protein Extraction

1. Dissect the tumor sample from the mice, trim away the fascia and membranous outer layer surrounding the tumor, and section the tumor into smaller pieces.
2. Lyse a small piece of tumor tissue in RIPA Lysis Buffer supplemented with protease inhibitor cocktail and grind it twice in a homogenizer for 30 s each at 15 s intervals (KZ-II, Servicebio).
3. Centrifuge at 16,000 *g* for 30 min at 4 °C.
4. Transfer the supernatant to a new tube and keep on ice.

3.4.2 Polyacrylamide Gel Electrophoresis (see **Note 9**)

1. Determine the protein concentration with a Bradford assay.
2. Mix the tissue extracts with loading buffer and denature in a 95 °C metal bath for 5 min. Cool at room temperature for 5 min and spin for 5 min. Store at –20 °C for future use.

3. Load equal amounts of protein and a molecular weight marker into the sodium dodecyl sulfate-polyacrylamide gel electrophoresis (SDS-PAGE) gels with appropriate concentration. Load 20–30 μg of total protein from tissue homogenate.
4. Run the gel for 1–2 h. It is recommended to use a voltage of 80 V for the stacking gel and 120 V for the separation gel.

*3.4.3 Western Blotting (see **Note 10**)*

1. Cut a nitrocellulose or PVDF membrane to the size of the gel (about 6 cm × 9 cm if using the Mini PROTEAN system). If using a PVDF membrane, activate it in methanol for about 30 s.
2. Incubate the membrane with transfer buffer (25 mM Tris, 192 mM glycine, 0.1% SDS, 20% methanol) until ready to use.
3. Transfer the protein from the gel to the nitrocellulose or PVDF membrane with variable power settings according to the manufacturer's instructions.

*3.4.4 Antibody Staining and Enhanced Chemiluminescence (see **Note 11**)*

1. Block the membrane with 5% non-fat milk in TBST (137 mM NaCl, 20 mM Tris, 0.1% Tween-20, pH 7.6) for 1 h at room temperature.
2. Dilute the primary antibody to the recommended dilution in 5% non-fat milk/TBST. Incubate the membrane in the primary antibody solution (at the appropriate dilution recommended in the product datasheet) overnight at 4 °C with gentle agitation.
3. Wash the membrane three times for 5 min each with 10 mL TBST.
4. Incubate the membrane with species-appropriate HRP-conjugated secondary antibody (1:1000–1:5000 in 5% non-fat milk/TBST) for 1 h with gentle agitation at room temperature.
5. Wash the membrane three times for 5 min each with 10 mL TBST.
6. Incubate the membrane with ECL (enhanced chemiluminescence substrate) for 1–2 min according to the manufacturer's detailed instructions.
7. Capture the chemiluminescence signals using digital imaging systems or X-ray film cassette.
8. Apply image analysis software to quantify band intensity and normalize target protein levels.

4 Notes

1. Glutaraldehyde is volatile, and osmium is highly toxic. They should be used in a fume hood to avoid exposure.

2. Use self-closing tweezers to hold the corner sides of the grid to avoid damaging the carbon coat.
3. Make sure the grids are thoroughly rinsed and completely dried before staining with lead citrate. Otherwise, a precipitate could form on the sections. Uranyl acetate is light-sensitive, and exposure to light should be limited.
4. Xylene is the most widely used clearing agent, but it is a toxic reagent. Less-toxic alternatives, such as xylene substitutes, are available.
5. Since the IHC protocol involves several harsh steps, adhesive-coated slides (such as silane or poly-L-lysine) or positively charged slides are required to avoid losing tissue from the glass slide.
6. Formalin crosslinks amino acids, resulting in the formation of methylene bridges that affect the epitope of the antigen. Therefore, an antigen-retrieval step is required to break these bridges and expose the antigenic sites so that the antibody can bind. Antigen retrieval can be achieved by either heat-induced epitope retrieval (HIER) or protease-induced epitope retrieval (PIER) methods.
7. Several tissues have endogenous peroxidase activity. If HRP label is being used for IHC, blocking of endogenous peroxidase by hydrogen peroxide treatment can be performed. Hydrogen peroxide suppresses endogenous peroxidase activity, thereby reducing background staining. Moreover, the need for hydrogen peroxide treatment is also determined by the tissue type since the endogenous expression of these enzymes and molecules varies among tissue types.
8. To obtain accurate experimental results, use fresh samples or store them at −80 °C (after snap freeze in liquid nitrogen) until use. Keep them on ice during the assay.
9. For large molecular proteins, the separation and migration in gel electrophoresis are slow, and the transfer from gel to membrane is also slow. Therefore, use a low concentration of gel, 8% or less. Because the glue in low concentration is fragile, it needs to be handled with great care.
10. Use tweezers to avoid finger contact with the membrane since the oil and protein on the finger can block the membrane and form dirty marks.
11. If necessary, the membrane could be stained with Ponceau S solution before the block to confirm protein transfer.

References

1. Dixon SJ, Lemberg KM, Lamprecht MR et al (2012) Ferroptosis: an iron-dependent form of nonapoptotic cell death. Cell 149:1060–1072
2. Tang D, Kroemer G (2020) Ferroptosis. Curr Biol 30:R1292–R12R7
3. Chen X, Yu C, Kang R et al (2021) Cellular degradation systems in ferroptosis. Cell Death Differ 28:1135–1148
4. Yang WS, SriRamaratnam R, Welsch ME et al (2014) Regulation of ferroptotic cancer cell death by GPX4. Cell 156:317–331
5. Liu J, Kang R, Tang D (2021) Signaling pathways and defense mechanisms of ferroptosis. FEBS J
6. Yang WS, Stockwell BR (2016) Ferroptosis: death by lipid peroxidation. Trends Cell Biol 26:165–176
7. Doll S, Proneth B, Tyurina YY et al (2017) ACSL4 dictates ferroptosis sensitivity by shaping cellular lipid composition. Nat Chem Biol 13:91–98
8. Yuan H, Li X, Zhang X et al (2016) Identification of ACSL4 as a biomarker and contributor of ferroptosis. Biochem Biophys Res Commun 478:1338–1343
9. Chen X, Comish PB, Tang D et al (2021) Characteristics and biomarkers of ferroptosis. Front Cell Dev Biol 9:637162
10. Chen X, Yu C, Kang R et al (2020) Iron metabolism in ferroptosis. Front Cell Dev Biol 8: 590226
11. Hou W, Xie Y, Song X et al (2016) Autophagy promotes ferroptosis by degradation of ferritin. Autophagy 12:1425–1428
12. Gao M, Monian P, Quadri N et al (2015) Glutaminolysis and transferrin regulate ferroptosis. Mol Cell 59:298–308
13. Wang Y, Liu Y, Liu J et al (2020) NEDD4L-mediated LTF protein degradation limits ferroptosis. Biochem Biophys Res Commun 531: 581–587
14. Chen X, Li J, Kang R et al (2021) Ferroptosis: machinery and regulation. Autophagy 17: 2054–2081
15. Feng H, Schorpp K, Jin J et al (2020) Transferrin receptor is a specific ferroptosis marker. Cell Rep 30:3411–3423 e7
16. Tu MJ, Yi CM, Traber GM et al (2022) Bioengineered RNA therapy in patient-derived organoids and xenograft mouse models. Methods Mol Biol 2521:191–206
17. Kuang F, Liu J, Tang D et al (2020) Oxidative damage and antioxidant defense in ferroptosis. Front Cell Dev Biol 8:586578

Chapter 18

Assessment of Ferroptosis in Hematopoietic Stem and Progenitor Cells

Yifan Zhang, Zhiyang Chen, Zhenyu Ju, and Qian Hu

Abstract

Hematopoietic stem cells (HSCs) are critical for maintaining hematopoiesis throughout life by utilizing their self-renewing and multipotent capabilities. Ferroptosis is a type of cell death characterized by the iron-dependent accumulation of lipid peroxides, and it is involved in multiple physiological and pathological conditions. Recent studies have highlighted the important role of ferroptosis in the functional maintenance of HSCs. Here, we describe our current protocols for accessing ferroptosis in hematopoietic stem and progenitor cells (HSPCs) both in vivo and in vitro. We introduce procedures for measuring total reactive oxygen species (ROS) and lipid ROS in HSPCs, as well as analyzing cell number, cell viability, and cell cycle profiles. This protocol provides a useful approach for characterizing the status of ferroptosis and its related parameters in HSPCs and more broadly, for studying the outcomes of ferroptosis on hematopoiesis.

Key words Hematopoietic stem and progenitor cell, Ferroptosis, Lipid peroxidation, Cell cycle

1 Introdution

As a research hotspot in the field of adult stem cells, the study of hematopoietic stem cells (HSCs) has important implications for the research of other stem cells. HSCs are at the top of the hematopoietic system, and in addition to their strong self-renewal ability, they differentiate into various progenitor cells, which further differentiate into various precursor cells and eventually develop into various mature cells in the blood system [1]. HSC transplantation is widely applied in the treatment of blood-related diseases such as bone marrow failure and leukemia [2].

Cell death plays an important role in the function and homeostasis of the hematopoietic system [3]. There have been a series of reports regarding the effects of apoptosis and necroptosis on HSCs [4–6]. Our recent report shows that hematopoietic stem and progenitor cells (HSPCs) are protected from ferroptosis by glutathione peroxidase 4 (GPX4) and Vitamin E in mice [7]. Furthermore,

Guido Kroemer and Daolin Tang (eds.), *Ferroptosis: Methods and Protocols*, Methods in Molecular Biology, vol. 2712,
https://doi.org/10.1007/978-1-0716-3433-2_18,

human HSCs have been reported to be selectively vulnerable to ferroptosis [8].

Understanding the relationship between HSCs and ferroptosis is beneficial for exploring the regulation of HSC function under physiological and pathological conditions. It also has significant reference value for studying the relationship between other stem cells and ferroptosis. In this chapter, we introduce a series of methods to detect HSC ferroptosis in mammals, both in vivo and in vitro.

2 Materials

The main materials are listed in Table 1.

Table 1
Reagent resource

Reagent or resource	Source	Identifier
Antibodies		
Anti-CD34-FITC	Biolegend	343510
Streptavidin-APC-Cy7	Thermo	47-4317-82
Anti-Flt3-PE	Thermo	12-1351-83
Anti-IL-7R-Percp-cy5.5	Biolegend	135022
Anti-CD16/32- AF700	Biolegend	56-0161-82
Anti-CD48- BV510	Biolegend	103443
Anti-CD150-BV605	Biolegend	115927
Anti-Sca-1-PE-Cy7	Biolegend	122514
Anti-C-kit-APC	Thermo	17-1171-83
Anti-Ki67-FITC	Biolegend	151204
Anti-CD11b-biotin	Biolegend	101204
Anti-Gr1-biotin	Biolegend	108404
Anti-Ter119-biotin	Biolegend	116204
Anti-B220-biotin	Biolegend	103204
Anti-CD4-biotin	Biolegend	100508
Anti-CD8-biotin	Biolegend	100704
Anti-APC MicroBeads	Miltenyi	130-090-855

(continued)

Table 1
(continued)

Reagent or resource	Source	Identifier
Chemicals, mediums, and commercial assays		
BIT9500	Stem cell	09500
L-glutamine	Gibco	25030081
β-ME	Amerisco	0482-250ML
Stem cell factor	PeproTech	250-03-100
Thrombopoietin	PeproTech	315-14-100
Murine IL-3	PeproTech	213-13-50
RBC lysis buffer (10×)	Biolengend	420301
Fixation/Permeabilization solution kit	BD	554714
C11-BODIPY	Invitrogen	D3861
DCFDA	Abcam	ab113851
DAPI	Sigma-Aldrich	D9542-10MG
DMEM/F12 medium	Gibco	11320033
Fetal bovine serum	ExCell	FSP500
Penicillin/streptomycin	Macklin	Q6532
Triton X100	Beyotime	P0096
PBS (powder)	Servicebio	G0002-2L
BSA	Sigma-Aldrich	V900933
75% ethanol		
Equipment		
Flow cell sorter	BD	BD Aria III
Flow cytometer	BD	BD Fortessa
Centrifuge	Eppendorf	5804
Pipettes	Eppendorf	2.5/10/20/200/1000 μL
96 U-bottom well plate	Corning	3599
Surgical scissors		
Teeth forceps		
Cell strainers		
Microcentrifuge tube	Axygen	1.5 mL
Mortar		
Cell culture incubator	Thermo fisher scientific	

(continued)

Table 1
(continued)

Reagent or resource	Source	Identifier
Centrifuge tube	Labselect	15 mL
Biological safety cabinets Clean benches	Thermo fisher scientific	1300 serials A2
LS separation columns	Miltenyi	130-042-401
MidiMACS separator	Miltenyi	130-042-302
5 mL polypropylene tube	Corning	352063

3 Flow Cytometry to Detect HSC Ferroptosis In Vivo

3.1 Overview

The proportion and number of each cell population in the hematopoietic hierarchy is an indicator of the homeostasis of the hematopoietic system. According to self-renewal ability and differentiation potential, HSCs are divided into long-term hematopoietic stem cells (LT-HSCs) and short-term hematopoietic stem cells (ST-HSCs). They differentiate into multipotent progenitors (MPPs), which then forms progenitor cells with specific differentiation tendency, including common lymphoid progenitors (CLPs), common myeloid progenitors (CMPs), and granulocyte-monocyte progenitors (GMPs) [1]. When HSPCs undergo ferroptosis, cell number and viability decrease along with an elevation of ROS and Lipid ROS level, normal or compensatory cell proliferation [9–11].

Different members in the HSPCs lineage are identified by flow cytometry according to their specific biological markers expressed on cell surface. The gating strategy is as follows: LT-HSC (Lineage$^-$Sca1$^+$c-Kit$^+$CD34$^+$Flt3$^-$CD48$^-$CD150$^+$), ST-HSC (Lineage$^-$Sca1$^+$c-Kit$^+$CD34$^-$Flt3$^-$CD48$^-$CD150$^-$), MPP4 (Lineage$^-$Sca1$^+$c-Kit$^+$CD34$^+$Flt3$^+$), CMP (Lineage$^-$Sca1$^-$c-Kit$^+$ CD34$^+$CD16/32$^-$), GMP (Lineage$^-$Sca1$^-$c-Kit$^+$CD34$^+$CD16/32$^+$), MEP (Lineage$^-$Sca1$^-$c-Kit$^+$CD34$^-$CD16/32$^-$), and CLP (Lineage$^-$Sca1$^-$c-Kit$^-$Flt-3$^+$IL-7R$^+$). Dead cells are labeled with DAPI. The viability and numbers of HSPCs are then analyzed.

As a nonhistone protein located in the nucleus, Ki-67 is expressed in all cell division phases except G0 phase [12]. In addition, cell DNA content varies from 2 N to 4 N along with the cell cycle. Thus, through DAPI/Ki-67 double staining, different phases of the cell cycle are identified as G0 phase (DAPI^{low}/Ki-67$^-$), G1 phase (DAPI^{low}/Ki-67$^+$), and S/G2/M phase (DAPI-^{high}/Ki-67$^+$) [12, 13].

The cell permeable dye 2′,7′-dichlorofluorescein diacetate (DCFDA) is used to measure hydroxy, hydrogen peroxide, and other ROS activities in cells. After DCFDA diffuses into cells, cellular esterase removes the DCFDA acetyl group, making it a non-fluorescent compound, which is then oxidized by ROS to 2′,7′-dichlorofluorescein (DCF). DCF is a strong fluorescent compound that can be detected by fluorescence spectroscopy, with maximum excitation wavelength and maximum emission wavelength of 495 nm and 529 nm [14].

3.2 The Measurement of the Number and Viability of HSPCs

1. Mice are sacrificed via cervical dislocation followed by a sterilization with 75% ethanol. All bones, including femurs, tibias, iliac bones, and spine, are isolated.
2. Remove the soft tissue on the bones with a sharp tip scissors and soft paper, place them in ice-cold PBS (*see* **Note 1**).
3. Transfer the bones to a mortar, add 3 mL ice-cold staining medium (0.2% BSA in PBS). Gently press the bones to release bone marrow cells (*see* **Note 2**).
4. Collect the suspension and filter it through a 200-mesh filter into a 15 mL centrifuge tube.
5. Repeat crushing the bone and collecting cell suspension until all bone marrow cells are released and the bones turn white.
6. Add ice-cold staining medium to adjust the cell suspension to 10 mL.
7. Prepare 1× RBC lysis buffer by diluting 10× RBC lysis buffer with ddH_2O.
8. After gentle vortex, 10 μL single cell suspension is transferred to a 1.5 mL microcentrifuge tube, mixed with 990 μL 1× RBC lysis buffer, and incubated at room temperature for 4–5 min with occasional shaking. Count the cell number (*see* **Note 3**).
9. Spin down remaining cell suspensions at 166× *g* for 5 min at 4 °C. Discard the supernatant, add staining medium to adjust cell concentration to 2×10^8 cell/ml .
10. Prepare Lineage antibody cocktail: mix 350 μL staining buffer with 100 μL Grl-Biotin, 100 μL Ter119-Biotin, 50 μL CD11b-Biotin, 50 μL B220-Biotin, 25 μL CD4-Biotin and 25 μL CD8-Biotin.
11. Transfer 50 μL (1×10^7) cells into a new 1.5 mL microcentrifuge tube. Add 4 μL of Lineage antibody cocktail. Incubate at 4 °C for 30 min.
12. Wash cells with 1 mL ice-cold staining medium. Centrifuge at 166× *g* for 5 min at 4 °C. Discard the supernatant.

13. Resuspend the pellet with 50 μL staining medium supplemented with antibodies containing 4 μL FITC-CD34, 4 μL APC-Cy7-Streptavidin, 1.8 μL PE-Flt3, 1.8 μL Percp-cy5.5-IL-7R, 1.8 μL AF-700-CD16/32, 1.8 μL BV510-CD48, 1.8 μL BV605-CD150, 1.5 μL PE-Cy7-Sca-1, and 1.5 μL APC-c-Kit. Incubate at 4 °C for 90 min in the dark.
14. Add 300 μL 1× RBC lysis buffer. Incubate at room temperature for 4–5 min in the dark with occasional shaking.
15. Add 1 mL ice-cold staining medium to stop red cell lysis, centrifuge at 166× *g* for 5 min at 4 °C. Discard the supernatant.
16. Add 500 μL ice-cold staining medium containing 1 μg/mL DAPI to resuspend cells and filter through a 300-mesh filter into a test tube. Cells are then ready to be analyzed with BD fortessa. The schematics of the HSPC populations gating strategy are given in Fig. 1.

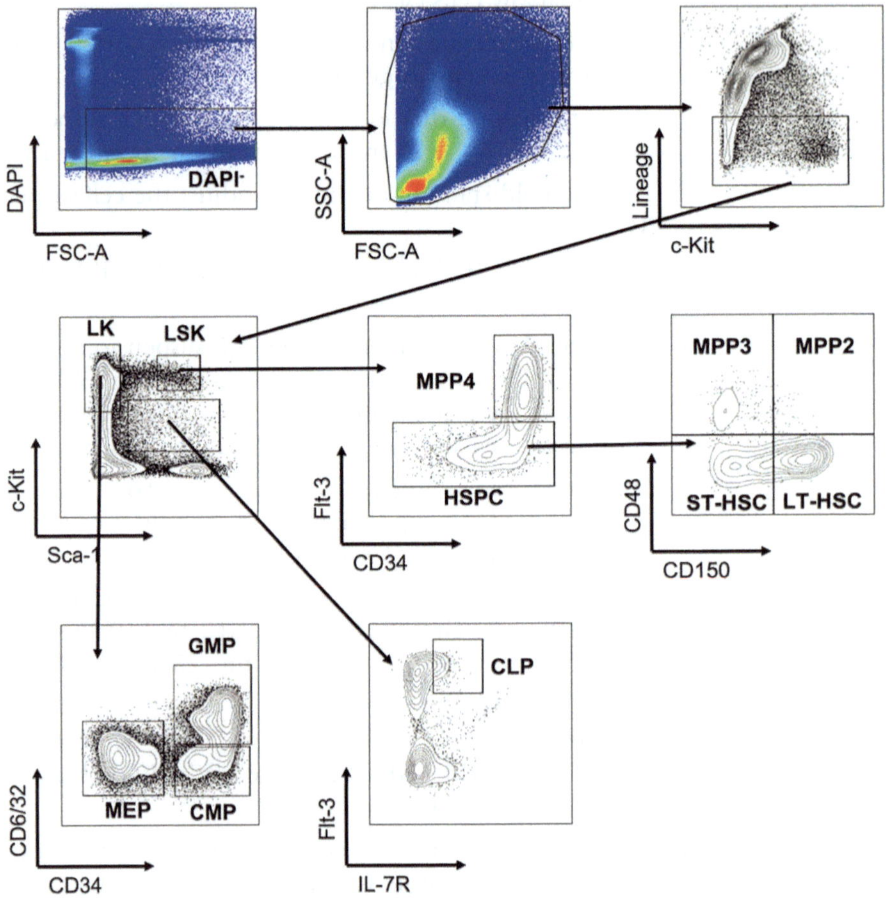

Fig. 1 Flow cytometry gating strategy of HSPCs in bone marrow of mice

3.3 The Measurement of the Cell Cycle of HSPCs

1. 1×10^7 bone marrow cells are isolated and stained with Lineage antibody cocktail following Subheading 3.2, **steps 1–12**.
2. Resuspend cell pellet by adding 50 μL staining medium supplemented with 4 μL APC-Cy7-Streptavidin, 1.8 μL BV510-CD48, 1.8 μL BV605-CD150, 1.5 μL PE-Cy7-Sca-1, 1.5 μL APC-c-Kit. Incubate at 4 °C for 30 min in dark (*see* **Note 4**).
3. Perform intracellular staining following a modified BD Cytofix/Cytoperm™ Fixation/Permeabilization Kit protocol (https://www.bdbiosciences.com) as below.
4. Add 120 μL BD Fixation/Permeabilization solution in the pellet, incubate at 4 °C for 20 min to fix cells.
5. Prepare 1× BD Perm/Wash Buffer through diluting 10× BD Perm/Wash Buffer with ddH_2O. Add 1 mL 1× BD Perm/Wash Buffer and spin at 166× *g* for 5 min at 4 °C. Discard the supernatant (*see* **Note 5**).
6. Gently resuspend cells with 100 μL 0.4% TritonX-100 (freshly made in PBS). Incubate for 1–2 min at room temperature for permeabilization.
7. Add 1 mL 1× BD Perm/Wash Buffer to clean cells, centrifuge at 166× *g* for 5 min at 4 °C, discard the supernatant.
8. Add 50 μL 1× BD Perm/Wash Buffer to resuspend cells, add 3 μL FITC-Ki67 antibody, incubate at room temperature for 40 min in the dark.
9. Add 1 mL 1× BD Perm/Wash Buffer to clean cells, centrifuge at 166× *g* for 5 min at 4 °C, discard the supernatant.
10. Add 500 μL ice-cold staining medium containing 20 μg/mL DAPI to resuspend cells and filter through a 300-mesh filter into a test tube. Cells are then ready to be analyzed with BD fortessa. The schematics of the HSPCs cell cycle gating strategy are given in Fig. 2.

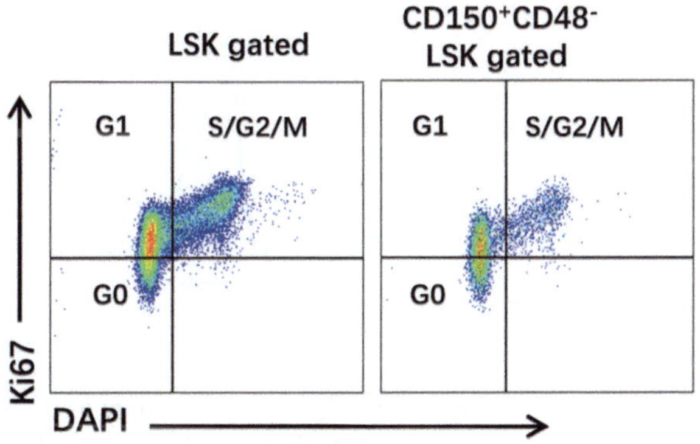

Fig. 2 Cell cycle gating strategy of HSPCs

3.4 ROS Measurements

A modified DCFDA/H2DCFDA-Cellular ROS Assay Kit protocol from Abcam (https://www.abcam.com/ab113851) is described in detail as follows:

1. 1×10^7 bone marrow cells are isolated and incubated with Lineage antibody cocktail following Subheading 3.2, **steps 1–12**.
2. Resuspend cell pellet by adding 50 μL staining medium supplemented with 4 μL APC-Cy7-Streptavidin, 1.8 μL BV510-CD48, 1.8 μL BV605-CD150, 1.5 μL PE-Cy7-Sca-1, 1.5 μL APC-c-Kit. Incubate at 4 °C for 30 min in the dark.
3. Add 300 μL 1× RBC lysis buffer and incubate at room temperature for 4–5 min in the dark with occasional shaking.
4. Add 1 mL ice-cold staining medium to stop red cell lysis, centrifuge at 166× *g* for 5 min at 4 °C. Discard the supernatant.
5. Make DCFDA work regent by diluting DCFDA with PBS to 2 μM.
6. Resuspend the cell pellet with 200 μL DCFDA work regent and incubate the cells for 30 min at 37 °C with occasional shaking.
7. Wash cells with 1 mL PBS, centrifuge at 166× *g* for 5 min at 4 ° C, discard the supernatant.
8. Repeat **step** 7 for 3 times.
9. Resuspend cells with 500 μL ice-cold staining medium containing 1 μg/mL DAPI and filter through a 300-mesh filter into a test tube. Cells are then ready to be analyzed with BD fortessa. The schematics of the HSPC ROS gating strategy are given in Fig. 3.

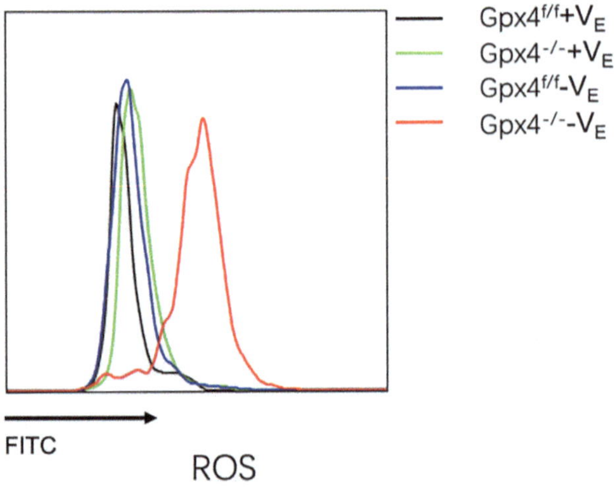

Fig. 3 The ROS gating strategy in LT-HSCs. Cells were gated from LT-HSCs in the *Gpx4 flox/flox* or *Gpx4 flox/flox* Vav Cre mice fed with a normal diet or with a V_E-deplete diet for 3 weeks. The *Gpx4* gene was specifically deleted in the hematopoietic system in the *Gpx4 flox/flox* Vav Cre mice

4 Detection of HSPC Ferroptosis In Vitro

4.1 Overview

Ferroptosis is a recently defined form of programmed cell death that is iron-dependent and is triggered by the accumulation of lethal levels of lipid peroxidation (Lipid ROS) [9]. Unlike apoptosis and necroptosis, ferroptosis is a distinct process characterized by unique molecular mechanisms and morphological features. To confirm the occurrence of ferroptosis, both iron-chelator (e.g., DFO and DFP) and lipophilic antioxidation (e.g., liproxstatin-1 and ferrostatin-1) should inhibit cell death. At the same time, the death of apoptosis and necroptosis should be ruled out by necroptosis inhibitors (Nec-1s) and apoptosis inhibitor (Z-VAD), respectively [15]. Furthermore, lipid ROS accumulation should occur during cell death. However, iron is necessary for the survival of HSPCs [16]. High concentration of iron chelator is toxic to HSPCs [7], which should be carefully considered when designing the experiments.

C11-BODIPY, a fluorescent probe that tracks Lipid ROS, can be oxidized by Lipid ROS in cells, resulting in a shift in emission wavelength from 591 nm to 510 nm. It is commonly used to measure Lipid ROS levels in ferroptosis cells [17].

4.2 Sorting and Culturing HSPCs In Vitro

1. Collect bone marrow cells following Subheading 3.2, **steps 1–5**. Spin cell suspension at 166× *g* for 5 min at 4 °C. Discard the supernatant.
2. Add 1 mL 1× Lysis Buffer, incubate at room temperature for 4–5 min with occasional shaking.
3. Add 1 mL ice-cold staining medium to stop lysis process, centrifuge at 166× *g* for 5 min at 4 °C. Discard the supernatant.
4. Isolate c-Kit$^+$ cells following a modified protocol from Milteny (https://www.miltenyibiotec.com) as follows.
5. Add 100 μL staining medium to resuspend cells and mix with c-Kit-APC antibody (2 μL per 10^8 bone marrow cells). Incubate at 4 °C for 30 min in dark.
6. Wash cells with 1 mL staining medium, centrifuge at 166× *g* for 5 min at 4 °C. Discard the supernatant.
7. Add 100 μL staining medium to resuspend cells and mix with anti-APC microbeads (5 μL per 10^8 bone marrow cells). Incubate at room temperature for 15 min in dark.
8. Wash cells with 1–2 mL ice-cold PBS, centrifuge at 166× *g* for 5 min at 4 °C. Discard the supernatant (*see* **Note 6**).
9. Resuspend cells in ice-cold PBS (500 μL per 10^8 bone marrow cells).
10. Place the LS column in the magnetic field of a suitable MACS Separator.

11. Add ice-cold PBS to rinse the LS column.
12. Slowly add the anti-APC microbeads-labeled cell suspension into the LS column through a 300-mesh filter. The magnetically labeled cells are retained within the column, while unlabeled cells flow through.
13. Collect unlabeled cells that pass through. Wash column for 3 times with 3 mL PBS. Collet the effluent (*see* **Note 7**).
14. Remove the LS column from the separator and place it on a new collection tube.
15. Pipette 3 mL ice-cold PBS into LS column. Immediately flush out fraction with the magnetically labeled cells by firmly applying the plunger supplied with the column.
16. Repeat **step 15** for 3 times. Centrifuge at 166× *g* for 5 min at 4 °C. Discard the c-Kit$^+$ cell supernatant.
17. Pipette 50 μL staining medium to resuspend cells, mix them with Lineage antibody cocktail. Incubate at 4 °C for 30 min.
18. Wash cells with 1 mL staining medium, centrifuge at 166× *g* for 5 min at 4 °C. Discard the supernatant.
19. Pipette 50 μL staining medium to resuspend cells, mix them with antibody cocktail (Refer to Subheading 3.2, **step 13** for staining). Incubate at 4 °C for 90 min in the dark.
20. Wash cells with 1 mL staining medium, centrifuge at 166× *g* for 5 min at 4 °C. Discard the supernatant.
21. Add 500 μL ice-cold staining medium containing 1 μg/mL DAPI to resuspend cells and filter through a 300-mesh filter into a test tube. Cells are then ready to be sorted with BD Arial III.
22. Prepare freshly made HSPC culture medium: DMEM/F12 medium supplemented with 10% fetal bovine serum, 20% BIT9500, 2 mM L-glutamine, 1% penicillin/streptomycin, 5×10^{-5} M β-ME, 10 ng/ml stem cell factor, 10 ng/ml thrombopoietin, and 15 ng/ml interleukin-3.
23. Resuspend cells with culture medium and transfer them into a 96 U-bottom well plate. Generally, we recommend culturing 5000 cells in 150 μL medium per well. Add 200 μL sterile PBS in the surrounding empty wells to minimize medium evaporation (*see* **Note 8**).
24. Culture cells in a 37 °C incubator with humidified atmosphere of 5% CO_2.

4.3 Lipid ROS Measurement

1. Cell ferroptosis are induced by 2 μM RSL3, and the control group is treated with the same volume of DMSO. The ferroptosis HSPCs are treated by Fer-1 (2 μM), Nec-1s (50 μM), or Z-VAD (50 μM).

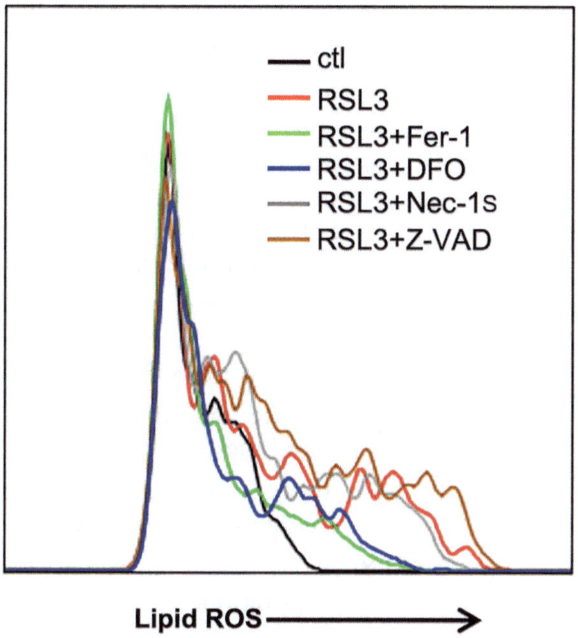

Fig. 4 The lipid ROS gating strategy in LSK cells. The cells were sorted and treated with ferroptosis inducer (RSL3) and inhibitor of ferroptosis, necroptosis, or apoptosis for 24 h

2. Test the level of lipid ROS in HSPCs after 24 h treatment.
3. Spin the plate at 166× *g* for 5 min. Remove the media carefully.
4. Add 100 μL 5 μM C11-BODIPY to each well, incubate at 37 °C for 30 min.
5. Centrifuge at 166× *g* for 5 min. Remove the supernatant.
6. Add 300 μL ice-cold staining medium containing 1 μg/mL DAPI to resuspend cells and filter through a 300-mesh filter into a test tube. The cells are then ready to be analyzed by BD fortessa (Fig. 4).

4.4 Cell Viability Measurement

1. Cell ferroptosis are induced by 2 μM RSL3. Test the viability of HSPCs after 48 h treatment.
2. Spin the plate at 166× *g* for 5 min. Remove the media carefully.
3. Resuspend cells with 300 μL ice-cold staining medium containing 1 μg/mL DAPI and filter through a 300-mesh filter into a test tube. Test the viability of HSPCs by flow cytometer.

5 Notes

1. The subsequent steps should be performed on ice unless specified, to minimize artificial disturbance.

2. Crushing instead of grinding to minimize cell death caused by manual manipulation.
3. About 2×10^8 bone marrow cells can be obtained from a 2-month-old C57 mouse.
4. It is recommended that the antibodies of cell surface markers are incubated at this step, to avoid some of them may not bind to fixed/denatured antigen. And we do not need to perform a red blood cell lysis step because BD Fixation/Permeabilization solution lyses red blood cells.
5. To keep cells permeabilized, they must be washed by BD Perm/Wash buffer in the following washing steps.
6. Replace staining buffer with PBS here to avoid the generation of bubbles, which may plug the column.
7. Always wait until the column reservoir is empty before proceeding to the next step.
8. Avoid using marginal wells as they are susceptible to evaporation.

References

1. Hofer T, Rodewald HR (2018) Differentiation-based model of hematopoietic stem cell functions and lineage pathways. Blood 132:1106–1113
2. Doulatov S, Notta F, Laurenti E et al (2012) Hematopoiesis: a human perspective. Cell Stem Cell 10:120–136
3. Tang D, Kang R, Berghe TV, et al (2019) The molecular machinery of regulated cell death. Cell Res 29(5):347–364
4. Roderick JE, Hermance N, Zelic M et al (2014) Hematopoietic RIPK1 deficiency results in bone marrow failure caused by apoptosis and RIPK3-mediated necroptosis. Proc Natl Acad Sci U S A 111:14436–14441
5. Yamashita M, Passegue E (2019) TNF-alpha coordinates hematopoietic stem cell survival and myeloid regeneration. Cell Stem Cell 25: 357–72 e7
6. Hu KX, Zhao SF, Guo M et al (2007) [Effects of mesenchymal stem cells on cell cycle and apoptosis of hematopoietic tissue cells in irradiated mice]. Zhongguo Shi Yan Xue Ye Xue Za Zhi 15:1226–1230
7. Hu Q, Zhang Y, Lou H et al (2021) GPX4 and vitamin E cooperatively protect hematopoietic stem and progenitor cells from lipid peroxidation and ferroptosis. Cell Death Dis 12:706
8. Zhao J, Jia Y, Mahmut D et al (2023) Human hematopoietic stem cell vulnerability to ferroptosis. Cell 186:732–747 e16
9. Xie Y, Hou W, Song X et al (2016) Ferroptosis: process and function. Cell Death Differ 23: 369–379
10. Dixon SJ, Lemberg KM, Lamprecht MR et al (2012) Ferroptosis: an iron-dependent form of nonapoptotic cell death. Cell 149:1060–1072
11. Tang D, Kroemer G (2020) Ferroptosis. Curr Biol 30:R1292–R12R7
12. Gerdes J, Lemke H, Baisch H et al (1984) Cell cycle analysis of a cell proliferation-associated human nuclear antigen defined by the monoclonal antibody Ki-67. J Immunol 133:1710–1715
13. Ormerod MG, Imrie PR (1990) Flow cytometry. Methods Mol Biol 5:543–558
14. Eruslanov E, Kusmartsev S (2010) Identification of ROS using oxidized DCFDA and flow-cytometry. Methods Mol Biol 594:57–72
15. Stockwell BR, Friedmann Angeli JP, Bayir H et al (2017) Ferroptosis: a regulated cell death nexus linking metabolism, redox biology, and disease. Cell 171:273–285
16. Zhang D, Gao X, Li H et al (2022) The microbiota regulates hematopoietic stem cell fate decisions by controlling iron availability in bone marrow. Cell Stem Cell 29:232–247 e7
17. Martinez AM, Kim A, Yang WS (2020) Detection of ferroptosis by BODIPY 581/591 C11. Methods Mol Biol 2108:125–130

Chapter 19

Detection of Ferroptosis by Immunohistochemistry and Immunofluorescence

Xiao Zhong and Ruochan Chen

Abstract

Ferroptosis is a type of regulated cell death driven by oxidative damage, characterized by iron overload and lipid peroxidation, and regulated by a network of distinct molecules and organelles. Impaired ferroptotic response is implicated in multiple physiological and pathological processes, including tumorigenesis, neurodegeneration, and ischemia-reperfusion damage. Classical techniques of immunohistochemistry (IHC) and immunofluorescence (IF) can be employed to exhibit antigen expression and location in tissues observed with microscopy, making them powerful tools in studying the ferroptosis process. In this chapter, we introduce commonly used protocols and summarize typical markers used in IHC and IF to monitor ferroptosis.

Key words Ferroptosis, Immunohistochemistry, Immunofluorescence

1 Introduction

Ferroptosis is a type of regulated cell death that is characterized by the accumulation of iron and lipid peroxidation, leading to membrane damage and cell death [1–3]. Unlike apoptosis, which involves programmed dismantling of cells, ferroptosis is an iron-dependent process that is regulated by a network of distinct molecular pathways and organelles [4, 5]. It has been implicated in multiple physiological and pathological processes, including tumorigenesis, neurodegeneration, and ischemia-reperfusion damage [6–8]. Understanding the mechanisms and biomarkers associated with ferroptosis is important for the development of potential therapeutic strategies to treat a wide range of diseases [4, 9, 10].

During ferroptosis, there is a shift in cellular metabolism toward increased lipid peroxidation and decreased antioxidant defense mechanisms, which is caused by alterations in the activity of several key enzymes and transporters involved in lipid

Guido Kroemer and Daolin Tang (eds.), *Ferroptosis: Methods and Protocols*, Methods in Molecular Biology, vol. 2712,
https://doi.org/10.1007/978-1-0716-3433-2_19,

metabolism [11]. Iron-dependent lipoxygenases (LOXs) catalyze the oxidation of polyunsaturated fatty acids, leading to the production of lipid hydroperoxides that can further react to generate toxic lipid radicals [12]. Malondialdehyde (MDA) and 4-hydroxy-nonenal (4-HNE) are popular biomarkers of ferroptosis [9]. Glutathione peroxidase 4 (GPX4) is an enzyme that can detoxify lipid hydroperoxides by reducing them to their corresponding alcohols [13]. However, in ferroptosis, the activity of GPX4 is suppressed, leading to an accumulation of lipid hydroperoxides and subsequent cell death [14]. Furthermore, the cystine/glutamate antiporter system xc^- is downregulated in ferroptosis, leading to a depletion of intracellular glutathione (GSH) [15]. GSH is an important antioxidant that can protect against oxidative stress, and its depletion further contributes to the accumulation of reactive oxygen species (ROS) in ferroptosis [16].

During ferroptosis, there is also a dysregulation of iron metabolism, leading to an increase in labile iron and a decrease in iron-binding proteins [17]. This leads to an accumulation of iron in the cytosol, where it can participate in the Fenton reaction and generate ROS, triggering ferroptotic cell death. Monitoring iron metabolism involved in iron uptake, storage, utilization, and release is a reasonable method to evaluate ferroptosis sensitivity [18]. Finally, the induction of ferroptosis is associated with an increased autophagic response to regulate protein or organelle degradation [19]. Multiple transcription factors, such as NRF2, TP53, and ATF4, regulate gene expression involved in iron and lipid metabolism, also determining the fate of ferroptosis [20].

Immunohistochemistry (IHC) and immunofluorescence (IF) are widely used imaging techniques for detecting proteins/antigens directly within tissues or cells, including in examinations of ferroptosis biomarkers. Unlike immunoblotting, which examines overall protein expression in tissues or cells, IHC and IF can localize the specific ferroptotic cellular distribution (e.g., membrane/cytoplasm/nucleus) and microenvironmental distribution, either from patient samples or animal tissues, by binding a specific antibody to the corresponding ferroptotic molecular target [21]. Both formalin-fixed paraffin-embedded (FFPE) tissues and fresh samples can be used for IHC and IF [22]. These methods are extensively used in diagnosis, localization, group classification, treatment, and prognosis in patient management, as well as in animal research [23–25]. In this chapter, we will present the methods of IHC and IF using examples of readily published ferroptotic regulators (Table 1).

Table 1
Immunological detection of ferroptosis-related proteins

Classified markers	Antibody target	Vendor (Catalog number)	Host/ Class	Reactivity	Applications	Publications
Iron metabolism	Ferritin	Abcam (ab75973)	Rabbit mAb	Human/ mouse/rat	IHC/IF	[37, 38]
	Transferrin	Abcam (ab82411)	Rabbit pAb	Mouse/rat/ Guinea/ pig/dog/ human	IHC/IF	[39, 40]
	TFR1	Invitrogen antibodies (13–6800)	Mouse pAb	Chicken/ hamster/ human/ mouse/rat	IHC/IF	[41, 42]
	Ferritin light chain	Abcam (ab69090)	Rabbit pAb	Human/ mouse/rat	IHC/IF	[43, 44]
	Ferritin heavy chain	Abcam (ab65080)	Rabbit pAb	Human	IHC/IF	[45, 46]
	STEAP3	Proteintech (17186–1-AP)	Rabbit pAb	Human/ mouse/rat	IHC/IF	[47, 48]
	DMT1	Proteintech (20507–1-AP)	Rabbit pAb	Human/ mouse/rat	IHC/IF	[49, 50]
	HO-1	Proteintech (10701–1-AP)	Rabbit pAb	Human/ mouse/rat	IHC/IF	[51, 52]
	IREB2	Abcam (ab181153)	Rabbit mAb	Human/ mouse/rat	IHC	[53]
	ACO1	Proteintech (12406–1-AP	Rabbit pAb	Human/ mouse/rat	IHC	[54]
	CISD1	Proteintech (16006–1-AP)	Rabbit pAb	Human/ mouse/rat	IHC/IF	[55]
	FTMT	Abcam (ab66111)	Rabbit pAb	Mouse/ human	IF	[56]
	PROM2	GeneTex (GTX83792)	Mouse mAb	Human/ mouse/ monkey	IHC/IF	[57]
Lipid metabolism	ACSL4	Proteintech (22401–1-AP)	Rabbit mAb	Human/ mouse/rat/ pig	IHC/IF	[58, 59]
	LPCAT3	Abcam (ab239585)	Mouse mAb	Human/pig	IHC	[60]
	ALOX5	Abcam (ab169755)	Rabbit mAb	Rat/human	IHC/IF	[61, 62]
	ALOX12	Abcam (ab167372)	Mouse pAb	Human	IHC/IF	[63, 64]
	ALOX15	Abcam (ab119774)	Mouse mAb	Human/ African green monkey	IHC/IF	[65, 66]
	ACSL1	Proteintech (13989–1-AP)	Rabbit pAb	Human/ mouse/rat	IHC/IF	[67, 68]
	PEBP1	MilliporeSigma (07–137)	Rabbit pAb	Human/ mouse/rat	IHC/IF	[69]

(continued)

Table 1
(continued)

Classified markers	Antibody target	Vendor (Catalog number)	Host/ Class	Reactivity	Applications	Publications
GSH metabolism	NRF2	Proteintech (16396–1-AP)	Rabbit pAb	Human/ mouse/rat	IHC/IF	[70]
	P53	Santa Cruz biotechnology (sc-126)	Mouse mAb	Human/ mouse/rat	IHC/IF	[71, 72]
	KEAP1	Proteintech (10503–2-AP)	Rabbit pAb	Human/ mouse/rat	IHC/IF	[73, 74]
	NOX1	Abcam (ab131088)	Rabbit pAb	Mouse/rat	IHC/IF	[75, 76]
	GPX4	Abcam (ab125066)	Rabbit mAb	Human/ mouse/rat	IHC/IF	[77, 78]
	SLC7A11	Cell Signaling Technology (12691)	Rabbit mAb	Human	IHC/IF	[79, 80]
	CDO1	Proteintech (12589–1-AP)	Rabbit pAb	Human/ mouse/rat/ pig	IHC	[81]
	GCLC	Proteintech (12601–1-AP)	Rabbit pAb	Human/ mouse/rat	IHC	[82]
	GCH1	Abcam (ab236387)	Mouse mAb	Human	IHC	[83]
Lipid peroxidation	4-HNE	Bioss (bs-6313R)	Rabbit pAb	Mouse/rat/ monkey	IHC/IF	[84, 85]
	MDA	Abcam (ab6463)	Rabbit pAb	Species independent	IHC/IF	[86, 87]

2 Materials

2.1 Tissue and Antibody Preparation

1. Tissue sections with tissue of interest (*see* **Note 1**).
2. Primary antibody specific for the target protein.
3. Secondary antibody that is reactive to primary antibody (*see* **Note 2**).

2.2 Immunohistochemistry Buffer (see Note 3)

1. Xylene.
2. Distilled H_2O (dH_2O).
3. 100% ethanol.
4. 95% ethanol: 95% ethanol in dH_2O.
5. 80% ethanol: 80% ethanol in dH_2O.
6. Sodium citrate solution:10 mM sodium citrate, 0.05% Tween 20, in dH_2O,pH 6.0.
7. Tris Buffered Saline with Tween 20 (TBST): 50 mM Tris, 150 mM NaCl, 0.05% Tween 20, in dH_2O, pH 7.6.

8. 3% Hydrogen Peroxide (H_2O_2): 3% H_2O_2 in dH_2O.
9. Blocking solution/antibody diluent: 5% goat serum in TBST.
10. 3,3′-diaminobenzidine (DAB) peroxidase substrate kit.

2.3 Immunofluorescence Buffer

1. 4% Formaldehyde (PFA): Methanol-Free.
2. Blocking Buffer: 5% normal serum, 0.3% Triton X-100 buffer in PBS.
3. Phosphate Buffered Saline (PBS).
4. Antibody diluent: 1% BSA, 0.3% Triton X-100 buffer in PBS.
5. Permeabilization buffer: 0.1% Triton X-100 in PBS.
6. DAPI solution.

2.4 Equipment

1. Staining holder.
2. Microwave.
3. Sharp-end tweezers.
4. PAP Pen.
5. Mounting medium.
6. Coverslips.
7. Humidified chamber protect from light.
8. Refrigerator.
9. Bright-field microscopy.
10. Fluorescent/confocal microscopy.

3 Methods

3.1 Monitoring Ferroptosis Using IHC

IHC is an important tool for studying the characteristics, localization, and relative quantification of antigens. The basic principle of IHC involves the specific binding of an immunological antibody-antigen complex, which is then visualized by a labeled antibody in tissue [26]. IHC was first introduced in the 1950s with the detection of antigens using antibodies with fluorescence, such as fluorescein isothiocyanate (FITC) [27, 28]. Nowadays, one of the most popular detecting methods for molecular targets is visualized with bright-field microscopy through enzymatic labels like horse radish peroxidase (HRP) and alkaline phosphatase (AP) [29].

Superior to IF, the bound antigen-antibody complex in IHC can be directly or indirectly visualized for a period at room temperature without signal quenching. The sensitivity and specificity of IHC are influenced by many variables in the procedure, which determine the positive and negative results of final imaging [21]. Appropriate antibody selection is vital for successfully identifying ferroptosis. IHC is still the most extensively used technique for clinical diagnosis in tissue samples [30].

In response to the increasing demand for detecting multiple signals simultaneously on a single tissue slide, fluorescent multiplex immunohistochemistry (mIHC) was developed [31, 32]. mIHC allows for comprehensive analysis of the co-expression of six or more ferroptotic markers, cell phenotype, antigen abundance, functional state, and cell-cell spatial relationship, providing many benefits, especially for samples with low availability [32]. The main technical principle of mIHC is based on Tyramide signal amplification (TSA) technology, which is an enzymatic assay that uses HRP for high-density in situ labeling of target proteins or nucleic acids [31]. The qualification and relative subcellular colocalization of ferroptosis markers SLC7A11/GPX4 and ACSL4/TFRC were validated in cervical cancer cell lines under anti-malignancy agents [33].

In this section, a modified IHC protocol (based on FFPE tissues) (https://www.cellsignal.com/learn-and-support/protocols/protocol-ihc-paraffin-signalstain) from Cell Signaling Technology is described in detail as follows:

1. Incubate tissue sections in xylene 3 times for 5 min each (*see* **Note 4**).
2. Incubate tissue sections in 100% ethanol 3 times for 10 min each (*see* **Note 5**).
3. Incubate tissue sections in 95% ethanol 2 times for 10 min each.
4. Incubate tissue sections in 80% ethanol 2 times for 10 min each.
5. Rinse tissue sections in dH_2O 2 times for 5 min each.
6. Antigen unmasking: Heat tissue sections in sodium citrate solution for 10 min at 95–98° followed by boiling initiated. Cool slides in sodium citrate solution on the bench until room temperature is reached (*see* **Note 6**).
7. Incubate tissue sections in 3 washes of dH_2O for 5 min each.
8. Incubate slides with 3% H_2O_2 for 10 min.
9. Immerse tissue sections in dH_2O 3 times for 5 min each.
10. Wash slides in TBST for 5 min.
11. Place each slide in 100–400 μL blocking solution for 1 h at room temperature (RT) (*see* **Note 7**).
12. Remove sections from blocking solution and cover tissue slides in primary antibody solution in antibody diluent (*see* **Note 8**) with equal volume as blocking solution overnight at 4 °C (*see* **Note 9**).
13. Remove primary antibody solution and wash sections in TBST 3 times for 5 min each.

14. Apply tissue slides with HRP-conjugated second antibody diluted in antibody diluent in recommended ratio and incubate for 30 min at RT in a humidified chamber.
15. Remove second antibody then rinse slides three times in TBST for 5 min each.
16. Remove TBST buffer and cover DAB substrate solutions with HRP compatible to the sections. Monitor the reaction closely and terminate this reaction as needed according to the manufacturer's instructions.
17. Rinse slides in dH_2O twice for 5 min each.
18. Incubate tissue sections in 80% ethanol twice for 10 s each.
19. Incubate tissue sections in 95% ethanol twice for 10 s each.
20. Incubate tissue sections in 100% ethanol twice for 10 s each.
21. Incubate tissue sections in xylene twice for 10 s each.
22. Apply sections with mounting medium and coverslips for imaging.

3.2 Monitoring Ferroptosis Using IF

As technology has developed and expanded, immunofluorescence (IF) has become the most extensively used tool in research, with multiple fluorescent indicators for colocalization [34]. Unlike IHC, the binding of fluorescence probes on the antibody and target must be incubated in a completely dark and humid chamber and observed with advanced equipment, such as fluorescent microscopy or confocal microscopy [35]. IF allows for the colocalization of three or more targets simultaneously or for restaining sequentially [36]. When dealing with IF, overlapping fluorescent signals should be avoided with an appropriate plan considering the species and fluorescent label selection, and autofluorescence in tissue or cells must be taken into consideration.

Here we introduce the IF method for the detection and localization of one target. A modified IF protocol (based on frozen sections and seeded cells) (https://www.cellsignal.com/learn-and-support/protocols/protocol-if) from Cell Signaling Technology is described in detail as follows:

1. Fix samples with 4% PFA for 15 min (*see* **Note 10**).
2. Incubate specimen in 3 washes of PBS for 5 min each.
3. Incubate specimen in permeabilization buffer for 30 min (*see* **Note 11**).
4. Immerse specimen with PBS 3 times for 5 min each.
5. Aspirate PBS and block specimen in blocking buffer for 60 min.

6. Remove blocking buffer and apply primary antibody diluted in antibody dilution buffer at a desired ratio to specimen.
7. Incubate specimen overnight at 4 °C.
8. Wash specimen in PBS 3 times for 5 min each.
9. Incubate specimen in fluorescent-labeled secondary antibody diluted in antibody dilution buffer along with DAPI solution for 1–2 h protected from light (*see* **Note 12**).
10. Wash specimen in PBS 3 times for 5 min each protected from light.
11. Airdry the samples then apply samples with mounting medium and coverslips for imaging.
12. Specimen should be stored at 4 °C protected from light for long-term storage.

4 Notes

1. FFPE and frozen sections can both be used for IHC and IF, but FFPE samples are more commonly used in IHC, while optimal fluorescence intensity is achieved in frozen slides using the IF method.
2. The secondary antibody used in IHC assay can be either HRP-conjugated or AP-conjugated enzymatic labels, depending on the substrate reaction.
3. Prepare solutions with reverse osmosis deionized or equivalent grade water.
4. Xylene is a flammable hazard. This procedure should be carried out in a chemical fume hood with outdoor exhaust ventilation.
5. Ethanol is a flammable hazard. This procedure should be carried out in a chemical fume hood with outdoor exhaust ventilation.
6. It is highly recommended to keep slides covered with buffer to maintain moisture on the surface of the tissue sample.
7. Circle around the sample on the glass slide with a PAP pen to create a water-proof barrier and reduce the volume of the solution.
8. Set antibody dilution gradient for optimal signal intensity. At the ideal dilution ratio, a significant difference in signal intensity should be observed between positive and negative samples, or consult the product-specific protocol for appropriate antibody dilution.
9. Make sure samples are completely covered with the solution.

10. All subsequent incubations should be carried out at room temperature (20–25 °C), unless noted otherwise.
11. If the target protein is located within the cell membrane, skip directly to **step 5**.
12. If a fluorochrome-conjugated primary antibody was used, counterstain samples with DAPI only.

References

1. Jiang X, Stockwell BR, Conrad M (2021) Ferroptosis: mechanisms, biology and role in disease. Nat Rev Mol Cell Biol 22:266–282
2. Stockwell BR (2022) Ferroptosis turns 10: emerging mechanisms, physiological functions, and therapeutic applications. Cell 185:2401–2421
3. Xie Y, Hou W, Song X et al (2016) Ferroptosis: process and function. Cell Death Differ 23: 369–379
4. Chen X, Li J, Kang R et al (2021) Ferroptosis: machinery and regulation. Autophagy 17: 2054–2081
5. Chen X, Kang R, Kroemer G et al (2021) Organelle-specific regulation of ferroptosis. Cell Death Differ 28:2843–2856
6. Tang D, Kroemer G, Kang R (2023) Ferroptosis in hepatocellular carcinoma: from bench to bedside. Hepatology. doi: 10.1097/HEP.0000000000000390. Publish Ahead of Print
7. Chen X, Kang R, Kroemer G et al (2021) Ferroptosis in infection, inflammation, and immunity. J Exp Med 218
8. Chen X, Kang R, Kroemer G et al (2021) Broadening horizons: the role of ferroptosis in cancer. Nat Rev Clin Oncol 18:280–296
9. Chen X, Comish PB, Tang D et al (2021) Characteristics and biomarkers of Ferroptosis. Front Cell Dev Biol 9:637162
10. Tang D, Chen X, Kang R et al (2021) Ferroptosis: molecular mechanisms and health implications. Cell Res 31:107–125
11. Liu J, Kang R, Tang D (2022) Signaling pathways and defense mechanisms of ferroptosis. FEBS J 289:7038–7050
12. Lin Z, Liu J, Kang R et al (2021) Lipid metabolism in ferroptosis. Adv Biol (Weinh) 5: e2100396
13. Yang WS, SriRamaratnam R, Welsch ME et al (2014) Regulation of ferroptotic cancer cell death by GPX4. Cell 156:317–331
14. Xue Q, Yan D, Chen X et al (2023) Copper-dependent autophagic degradation of GPX4 drives ferroptosis. Autophagy:1–15
15. Badgley MA, Kremer DM, Maurer HC et al (2020) Cysteine depletion induces pancreatic tumor ferroptosis in mice. Science 368:85–89
16. Liu Y, Wang Y, Lin Z et al (2023) SLC25A22 as a key mitochondrial transporter against ferroptosis by producing GSH and MUFAs. Antioxid Redox Signal
17. Torti SV, Torti FM (2020) Iron: the cancer connection. Mol Asp Med 75:100860
18. Chen X, Yu C, Kang R et al (2020) Iron metabolism in Ferroptosis. Front Cell Dev Biol 8: 590226
19. Chen F, Cai X, Kang R et al (2023) Autophagy-dependent Ferroptosis in cancer. Antioxid Redox Signal
20. Dai C, Chen X, Li J et al (2020) Transcription factors in ferroptotic cell death. Cancer Gene Ther 27:645–656
21. Hussaini HM, Seo B, Rich AM (2023) Immunohistochemistry and immunofluorescence. Methods Mol Biol 2588:439–450
22. Shi SR, Shi Y, Taylor CR (2011) Antigen retrieval immunohistochemistry: review and future prospects in research and diagnosis over two decades. J Histochem Cytochem 59: 13–32
23. Van Der Kraak L, Goel G, Ramanan K, et al (2016). 5-Fluorouracil upregulates cell surface B7-H1 (PD-L1) expression in gastrointestinal cancers. J Immunother Cancer 18 (4):65
24. Weis CA, Kather JN, Melchers S et al (2018) Automatic evaluation of tumor budding in immunohistochemically stained colorectal carcinomas and correlation to clinical outcome. Diagn Pathol 13:64
25. Boxberg M, Gotz C, Haidari S et al (2018) Immunohistochemical expression of CD44 in oral squamous cell carcinoma in relation to histomorphological parameters and clinicopathological factors. Histopathology 73:559–572
26. Duraiyan J, Govindarajan R, Kaliyappan K et al (2012) Applications of immunohistochemistry. J Pharm Bioallied Sci 4:S307–S309

27. Coons AH, Leduc EH, Kaplan MH (1951) Localization of antigen in tissue cells. VI. The fate of injected foreign proteins in the mouse. J Exp Med 93:173–188
28. Coons AH, Kaplan MH (1950) Localization of antigen in tissue cells; improvements in a method for the detection of antigen by means of fluorescent antibody. J Exp Med 91:1–13
29. Farr AG, Nakane PK (1981) Immunohistochemistry with enzyme labeled antibodies: a brief review. J Immunol Methods 47:129–144
30. Idikio HA (2009) Immunohistochemistry in diagnostic surgical pathology: contributions of protein life-cycle, use of evidence-based methods and data normalization on interpretation of immunohistochemical stains. Int J Clin Exp Pathol 3:169–176
31. Tan WCC, Nerurkar SN, Cai HY et al (2020) Overview of multiplex immunohistochemistry/immunofluorescence techniques in the era of cancer immunotherapy. Cancer Commun (Lond) 40:135–153
32. Stack EC, Wang C, Roman KA et al (2014) Multiplexed immunohistochemistry, imaging, and quantitation: a review, with an assessment of Tyramide signal amplification, multispectral imaging and multiplex analysis. Methods 70: 46–58
33. Zhao MY, Liu P, Sun C et al (2022) Propofol augments paclitaxel-induced cervical cancer cell Ferroptosis in vitro. Front Pharmacol 13: 816432
34. Taraska JW, Zagotta WN (2010) Fluorescence applications in molecular neurobiology. Neuron 66:170–189
35. Konishi H, Steinbach G, Hittelman WN et al (1996) Cell kinetic analysis of intact rat colonic crypts by confocal microscopy and immunofluorescence. Gastroenterology 111:1493–1500
36. van de Werken C, Jahr H, Avo Santos M et al (2013) A universal method for sequential immunofluorescent analysis of chromatin and chromatin-associated proteins on chromosome spreads. Chromosom Res 21:475–489
37. Belaya I, Kucharikova N, Gorova V et al (2021) Regular physical exercise modulates iron homeostasis in the 5xFAD mouse model of Alzheimer's disease. Int J Mol Sci 22
38. Ariza J, Steward C, Rueckert F et al (2015) Dysregulated iron metabolism in the choroid plexus in fragile X-associated tremor/ataxia syndrome. Brain Res 1598:88–96
39. Mamais A, Kluss JH, Bonet-Ponce L et al (2021) Mutations in LRRK2 linked to Parkinson disease sequester Rab8a to damaged lysosomes and regulate transferrin-mediated iron uptake in microglia. PLoS Biol 19:e3001480
40. Yang AC, Stevens MY, Chen MB et al (2020) Physiological blood-brain transport is impaired with age by a shift in transcytosis. Nature 583: 425–430
41. Chen S, Li W, Zheng X et al (2022) Immunomagnetic microscopy of tumor tissues using quantum sensors in diamond. Proc Natl Acad Sci U S A 119
42. Vermeulen JF, van Brussel AS, van der Groep P et al (2012) Immunophenotyping invasive breast cancer: paving the road for molecular imaging. BMC Cancer 12:240
43. Wang Y, Qiu S, Wang H et al (2021) Transcriptional repression of ferritin light chain increases Ferroptosis sensitivity in lung adenocarcinoma. Front Cell Dev Biol 9:719187
44. Chandran V, Gao K, Swarup V et al (2017) Inducible and reversible phenotypes in a novel mouse model of Friedreich's ataxia. elife 6
45. You L, Yu PP, Dong T et al (2022) Astrocyte-derived hepcidin controls iron traffic at the blood-brain-barrier via regulating ferroportin 1 of microvascular endothelial cells. Cell Death Dis 13:667
46. van Raaij S, van Swelm R, Bouman K et al (2018) Tubular iron deposition and iron handling proteins in human healthy kidney and chronic kidney disease. Sci Rep 8:9353
47. Wang S, Chen L, Liu W (2022) Matrix stiffness-dependent STEAP3 coordinated with PD-L2 identify tumor responding to sorafenib treatment in hepatocellular carcinoma. Cancer Cell Int 22:318
48. Zhu Y, Peng X, Zhou Q et al (2022) METTL3-mediated m6A modification of STEAP2 mRNA inhibits papillary thyroid cancer progress by blocking the hedgehog signaling pathway and epithelial-to-mesenchymal transition. Cell Death Dis 13:358
49. Yamaguchi H, Honda S, Torii S et al (2020) Wipi3 is essential for alternative autophagy and its loss causes neurodegeneration. Nat Commun 11:5311
50. Song X, Zhu S, Chen P et al (2018) AMPK-mediated BECN1 phosphorylation promotes Ferroptosis by directly blocking system X(c)(−) activity. Curr Biol 28:2388–99. e5, 2388
51. Chen H, Li Z, Xu J et al (2023) Curcumin induces Ferroptosis in follicular thyroid cancer by upregulating HO-1 expression. Oxidative Med Cell Longev 2023:6896790
52. Wang S, Zheng Y, Jin S et al (2022) Dioscin protects against cisplatin-induced acute kidney injury by reducing Ferroptosis and apoptosis through activating Nrf2/HO-1 signaling. Antioxidants (Basel) 11

53. Cloonan SM, Glass K, Laucho-Contreras ME et al (2016) Mitochondrial iron chelation ameliorates cigarette smoke-induced bronchitis and emphysema in mice. Nat Med 22:163–174
54. Niu L, Ye C, Sun Y et al (2018) Mutant huntingtin induces iron overload via up-regulating IRP1 in Huntington's disease. Cell Biosci 8:41
55. Geldenhuys WJ, Benkovic SA, Lin L et al (2017) MitoNEET (CISD1) knockout mice show signs of striatal mitochondrial dysfunction and a Parkinson's disease phenotype. ACS Chem Neurosci 8:2759–2765
56. Gong S, Chen Y, Meng F et al (2017) Roflumilast restores cAMP/PKA/CREB signaling axis for FtMt-mediated tumor inhibition of ovarian cancer. Oncotarget 8:112341–112353
57. Ziller N, Kotolloshi R, Esmaeili M et al (2020) Sex differences in diabetes- and TGF-beta1-induced renal damage. Cell 9
58. Zhou X, Cao T (2022) Zinc oxide nanoparticle inhibits tumorigenesis of renal cell carcinoma by modulating lipid metabolism targeting miR-454-3p to repressing metabolism enzyme ACSL4. J Oncol 2022:2883404
59. Liu C, Sun W, Zhu T et al (2022) Glia maturation factor-beta induces ferroptosis by impairing chaperone-mediated autophagic degradation of ACSL4 in early diabetic retinopathy. Redox Biol 52:102292
60. Wei J, Nai GY, Dai Y et al (2021) Dipetidyl peptidase-4 and transferrin receptor serve as prognostic biomarkers for acute myeloid leukemia. Ann Transl Med 9:1381
61. Tomita K, Takashi Y, Ouchi Y et al (2019) Lipid peroxidation increases hydrogen peroxide permeability leading to cell death in cancer cell lines that lack mtDNA. Cancer Sci 110: 2856–2866
62. Rodriguez-Blanco G, Zeneyedpour L, Duijvesz D et al (2018) Tissue proteomics outlines AGR2 AND LOX5 as markers for biochemical recurrence of prostate cancer. Oncotarget 9: 36444–36456
63. Kim JY, Choi BG, Jelinek J et al (2020) Promoter methylation changes in ALOX12 and AIRE1: novel epigenetic markers for atherosclerosis. Clin Epigenetics 12:66
64. Zhang Q, Yan G, Lei J et al (2020) The SP1-12LOX axis promotes chemoresistance and metastasis of ovarian cancer. Mol Med 26: 39
65. Khophai S, Thanee M, Techasen A et al (2018) Zileuton suppresses cholangiocarcinoma cell proliferation and migration through inhibition of the Akt signaling pathway. Onco Targets Ther 11:7019–7029
66. Werz O, Gerstmeier J, Libreros S et al (2018) Human macrophages differentially produce specific resolvin or leukotriene signals that depend on bacterial pathogenicity. Nat Commun 9:59
67. Liu F, Tang L, Li Q et al (2022) Single-cell transcriptomics uncover the key ferroptosis regulators contribute to cancer progression in head and neck squamous cell carcinoma. Front Mol Biosci 9:962742
68. Wang Q, Cai J, Fang C et al (2018) Mesenchymal glioblastoma constitutes a major ceRNA signature in the TGF-beta pathway. Theranostics 8:4733–4749
69. Martinho O, Pinto F, Granja S et al (2013) RKIP inhibition in cervical cancer is associated with higher tumor aggressive behavior and resistance to cisplatin therapy. PLoS One 8: e59104
70. Sheng Z, Zheng F, Li J et al (2022) Denervation delays initial bone healing of rat tooth extraction socket. Odontology 111:640
71. Oturkar CC, Gandhi N, Rao P et al (2022) Estrogen receptor-Beta2 (ERbeta2)-mutant p53-FOXM1 Axis: a novel driver of proliferation, Chemoresistance, and disease progression in high grade serous ovarian cancer (HGSOC). Cancers (Basel) 14
72. Calabrese C, Panuzzo C, Stanga S et al (2020) Deferasirox-dependent iron chelation enhances mitochondrial dysfunction and restores p53 signaling by stabilization of p53 family members in leukemic cells. Int J Mol Sci 21
73. Sargin P, Roethle MF, Jia S et al (2022) Lactiplantibacillus plantarum 299v supplementation modulates beta-cell ER stress and antioxidative defense pathways and prevents type 1 diabetes in gluten-free BioBreeding rats. Gut Microbes 14:2136467
74. Pei J, Wu M, Cai S et al (2022) The protective effect of Ursolic acid on unilateral ureteral obstruction in rats by activating the Nrf2/HO-1 antioxidant signaling pathway. Comput Intell Neurosci 2022:3690524
75. Zhu N, Chen X, Zhao J et al (2022) Hypoxia-induced LINC00674 facilitates hepatocellular carcinoma progression by activating the NOX1/mTOR signaling pathway. J Cancer 13:3177–3188
76. Xia M, Chen W, Wang J et al (2019) TRPA1 activation-induced myelin degradation plays a key role in motor dysfunction after intracerebral hemorrhage. Front Mol Neurosci 12:98
77. Ma X, Dong X, Xu Y et al (2022) Identification of AP-1 as a critical regulator of glutathione peroxidase 4 (GPX4) transcriptional

suppression and acinar cell Ferroptosis in acute pancreatitis. Antioxidants (Basel) 12
78. Mayr L, Grabherr F, Schwarzler J et al (2020) Dietary lipids fuel GPX4-restricted enteritis resembling Crohn's disease. Nat Commun 11:1775
79. Zhao YY, Yang YQ, Sheng HH et al (2022) GPX4 plays a crucial role in Fuzheng Kang'ai decoction-induced non-small cell lung cancer cell Ferroptosis. Front Pharmacol 13:851680
80. Zhang Y, Shi J, Liu X et al (2018) BAP1 links metabolic regulation of ferroptosis to tumour suppression. Nat Cell Biol 20:1181–1192
81. Luo W, Dai W, Li Q et al (2022) Ferroptosis-associated molecular classification characterized by distinct tumor microenvironment profiles in colorectal cancer. Int J Biol Sci 18: 1773–1794
82. Hu X, Mu L, Zhu L et al (2021) Lycium barbarum polysaccharides attenuate cardiovascular oxidative stress injury by enhancing the Keap1/Nrf2 signaling pathway in exhaustive exercise rats. Mol Med Rep 24
83. Chen T, Yang H, Liu P et al (2022) Splicing factor SF3B3, a NS5-binding protein, restricts ZIKV infection by targeting GCH1, vol 38, Virol Sin, p 222
84. Pyun BJ, Jo K, Lee JY et al (2022) Caesalpinia sappan Linn. Ameliorates allergic nasal inflammation by upregulating the Keap1/Nrf2/HO-1 pathway in an allergic rhinitis mouse model and nasal epithelial cells. Antioxidants (Basel) 11
85. Polverino F, Doyle-Eisele M, McDonald J et al (2015) A novel nonhuman primate model of cigarette smoke-induced airway disease. Am J Pathol 185:741–755
86. Napso T, Lean SC, Lu M et al (2022) Diet-induced maternal obesity impacts feto-placental growth and induces sex-specific alterations in placental morphology, mitochondrial bioenergetics, dynamics, lipid metabolism and oxidative stress in mice. Acta Physiol (Oxf) 234:e13795
87. Ide S, Kobayashi Y, Ide K et al (2021) Ferroptotic stress promotes the accumulation of pro-inflammatory proximal tubular cells in maladaptive renal repair. elife 10

Chapter 20

Analysis of MicroRNAs in Ferroptosis

John G. Yuen, Ga-Ram Hwang, and Jingfang Ju

Abstract

MicroRNAs (miRNAs) are a class of small non-coding RNA molecules that are involved in a wide range of biological processes, including development, differentiation, and disease. They function by binding to the 3′ untranslated region (UTR) of target mRNAs, leading to mRNA degradation or translational repression. miRNAs are involved in the regulation of many cellular processes, including cell proliferation, apoptosis, and metabolism. MiRNAs have been shown to modulate ferroptosis in several ways. Some miRNAs have been shown to promote ferroptosis by increasing the expression of genes involved in lipid peroxidation. Other miRNAs have been shown to inhibit ferroptosis by decreasing the expression of genes involved in iron uptake. The role of miRNAs in ferroptosis is still being studied, but they play a significant role in this cell death pathway. miRNAs may be potential targets for therapeutic intervention in diseases associated with ferroptosis, such as cancer and neurodegenerative diseases. This chapter outlines several methods used to study the connection between miRNAs and ferroptosis through target discovery and validation.

Key words MicroRNAs, Ferroptosis

1 Introduction

MicroRNAs (miRNAs) are a class of small non-coding RNAs that play an important role in regulating gene expression via the RNA-induced silencing (RNAi) pathway [1–3]. They are typically 20–25 nucleotides long and bind to the 3′ untranslated region (UTR) on messenger RNAs (mRNAs), which can lead to the inhibition of translation or the degradation of the mRNA. This binding interaction is largely mediated by the seed sequence, located on positions 2–7 of the miRNA, but is also influenced by potential base pairing on the 3′ end of the miRNA [4–6]. Due to their integral role in influencing protein expression, their dysregulation impacts many biological processes and has been linked to a number of diseases, including cancer, cardiovascular disease, and neurodegenerative diseases [7–10]. Ferroptosis has been found to be associated with these diseases as well, and several studies have found that miRNAs can also regulate ferroptosis

Guido Kroemer and Daolin Tang (eds.), *Ferroptosis: Methods and Protocols*, Methods in Molecular Biology, vol. 2712, https://doi.org/10.1007/978-1-0716-3433-2_20,

[11, 12]. Therefore, studying the impact of miRNAs on the regulation of ferroptosis may yield promising results in the study of these diseases.

Recent studies have found that miRNAs can regulate ferroptosis through several pathways, including iron metabolism, lipid metabolism, and antioxidant defense pathways. For example, miR-17 has been shown to regulate ferroptosis by targeting the iron exporter protein ferroportin (FPN1), which is involved in the export of iron from cells [13]. Certain miRNAs also regulate upstream targets, for example, miR-137 inhibits solute carrier family 1 member 5 (SLC1A5, also known as ASCT2), a neutral amino acid transporter that is involved in glutamine homeostasis [14]. Dysregulation of glutamine homeostasis can lead to cysteine starvation and therefore downstream glutathione depletion [15]. Other miRNAs, such as miR-1287 and miR-107, have been shown to regulate ferroptosis by targeting the glutathione peroxidase 4 (GPX4), a key enzyme in the antioxidant defense against lipid peroxidation that inhibits ferroptosis [16]. In prostate and colon cancer, miR-15a was found to target GPX4, and overexpression of miR-15a resulted in promoting ferroptosis [17, 18]. Solute carrier family 7 member 11 (SLC7A11, also known as xCT) has also been shown to inhibit ferroptosis, and recent studies have demonstrated that miRNAs target SLC7A11 [15, 19]. For example, miR-375 was found to target SLC7A11, and overexpression of miR-375 was found to promote ferroptosis [20]. Overall, miRNAs impact ferroptosis via several different mechanisms and therefore, dysregulated expression of miRNAs may influence ferroptosis in the pathophysiology of diseases.

The regulation of ferroptosis by miRNAs is a complex and rapidly evolving area of research with important implications for understanding the molecular mechanisms underlying various diseases and developing new therapeutic strategies. Modifying the expression of miRNAs has been an area of therapeutic focus, particularly on restoring the expression of downregulated miRNAs in cancer [21–23]. While there are still hurdles to overcome in the development and delivery of miRNAs as a therapeutic, inducing ferroptosis may be an avenue to consider in the development of miRNA-based therapeutics. There are a number of well-established technologies to quantify miRNA expression and to identify direct targets of miRNA [24, 25]. This chapter provides an overview of several widely used methods for investigating the miRNAs that may be involved in regulating ferroptosis.

2 Materials

1. Ethanol, Absolute (200 Proof), Molecular Biology Grade (Fisher Scientific, BP2818).

2. Isopropanol, Absolute (200 Proof), Molecular Biology Grade (Fisher Scientific, BP2618).
3. DPBS, No Calcium, No Magnesium (Thermo Fisher, 14190250).
4. TRIzol™ Reagent (Thermo Fisher, 15596018).
5. BCP Phase Separation Reagent (Molecular Research Center, BP 151).
6. Nuclease-free water (Not DEPC-Treated) (Thermo Fisher, AM9937).
7. High-Capacity cDNA Reverse Transcription Kit with RNAse Inhibitor, 1000 Reactions (Thermo Fisher, 4374967). Kit includes the following components:10× RT Buffer; 10× RT Random Primers; 25X dNTP Mix (100 mM); MultiScribe™ Reverse Transcriptase (50 U/μl); and RNase Inhibitor.
8. TaqMan™ Universal PCR Master Mix, no AmpErase™ UNG, 200 Reactions (Thermo Fisher, 4324018).
9. TaqMan™ MicroRNA Assay (Thermo Fisher). PCR and RT-PCR probes are inventoried and made to order based on miRNA of interest.
10. Pre-miR™ miRNA Precursor Molecules – Negative Control #2 (Thermo Fisher, AM17111).
11. miRNA mimics/anti-miRs (can be purchased from vendors such as Thermo Fisher and Horizon Discovery Biosciences).
12. Oligofectamine™ Transfection Reagent (Thermo Fisher, 1225011).
13. Tissue Culture Media (Dependent on cell line being used, i.e., DMEM).
14. Fetal Bovine Serum, Heat Inactivated, Sterile-filtered (Millipore-Sigma, F4135-500ML).
15. Opti-MEM™ Reduced Serum Medium (Thermo Fisher, 31985070).
16. pGL4 Vector (Addgene 48744).
17. pR L-SV40 Reporter Vector (Promega, E2231),
18. DharmaFECT™ Duo Transfection Reagent (Horizon Discovery, T-2010-01).
19. Dual-Luciferase® Reporter Assay System (Promega, E1910).
20. RNaseZap™ Rnase Decontamination Solution (Thermo Fisher, AM9780).
21. Falcon 96-Well Cell Culture Plates, Corning, Black/Clear Bottom (VWR, 89022-352).
22. Falcon 96-Well Cell Culture Plates, Corning, Clear (VWR, 62406-081).

23. 96-Well PCR μltraPlate (Sorenson BioSciences, 21970).
24. Falcon 6-Well Cell Culture Plates, Corning (VWR, 62406-161).
25. MicroAmp™ Clear Adhesive Film (Thermo Fisher, 4306311).
26. Nanodrop™ 2000 Spectrophotometer (Thermo Fisher, ND-2000).
27. MiniAmp™ Plus Thermal Cycler (Thermo Fisher, A37835).
28. QuantStudio™ 3 Real-Time PCR System, 96-Well, 0.2 ml (Thermo Fisher, A28567).
29. SpectraMax iD3 (Molecular Devices).
30. Eppendorf Centrifuge 5424 R (Eppendorf).

3 Methods

3.1 Discovering and Screening for miRNAs Involved in Ferroptosis with RT-PCR or MicroRNA Sequencing

Screening for miRNAs is a common approach to identify miRNAs involved in a particular biological and/or disease process. Real-time PCR (RT-PCR) of miRNAs is a commonly used tool to detect levels of miRNAs in tissues and cells and can be scaled to screen a moderate number of miRNAs. This technique is particularly useful when assessing the levels of miRNAs when there are specific candidate miRNAs to evaluate. MiRNA sequencing, on the other hand, is a more unbiased approach, revealing the expression of all miRNAs in the tissues/cells of interest. Both techniques involve extracting total RNA from the sample and then subsequently processing them for RT-PCR or sequencing. RT-PCR involves a reverse transcription step, followed by a quantitative real-time PCR reaction. MiRNA sequencing typically employs adapter ligation step(s) prior to next-generation sequencing [26, 27].

Both forward and reverse genetics approaches can be taken to investigate miRNAs involved in ferroptosis. Key genes involved in the regulation of ferroptosis can be selected and in silico miRNA target prediction tools, such as TargetScan or miRanda, can be used to identify candidate miRNAs that may bind to these gene(s) of interest [28, 29]. These candidate miRNAs can then be quantified with techniques such as RT-PCR. With a forward genetics approach, ferroptosis can be inhibited or induced in conditions/diseases of interest and changes in the relative expression of different miRNAs can be evaluated as potential regulators of ferroptosis.

3.1.1 Isolation of Total RNA

1. Plate cells at appropriate density and incubate at 37 °C overnight. For RT-PCR, around 1.0×10^6–2.0×10^6 cells plated onto a 6-well plate is typical. More cells may be desired for RNAseq.

2. Prepare 75% ethanol and keep on ice.
3. Aspirate media from sample(s) on a 6-well plate and wash with PBS.
4. Add 1 mL TRIzol reagent to each sample and incubate for 2–3 min. Pipette up and down or scrape sample before transferring into a 1.7 mL tube.
5. Add 100 μL BCP to the tube and vortex vigorously.
6. Incubate sample at room temperature for 2–3 min.
7. Centrifuge for 15 min at 4 °C at full speed (13,200 rpm).
8. Transfer the top aqueous layer to a new tube.
9. Add 0.5 mL of isopropanol to each tube, vortex, and incubate for 30 min at −80 °C.
10. Centrifuge for 10 min at 4 °C at full speed (13,200 rpm).
11. Carefully remove all the supernatant while preserving the white pellet at the bottom of the tube.
12. Wash pellet with 1 mL of 75% Ethanol.
13. Centrifuge for 10 min at 4 °C at full speed (13,200 rpm) and carefully remover supernatant.
14. Briefly air-dry pellet and resuspend in nuclease-free water.
15. Dilute 1 μL of RNA in 9 μL of nuclease-free water and measure RNA concentration using the Nanodrop machine. Also, be sure to measure the optical density (O.D.) at 260 nm and 280 nm to assess RNA quality and purity.
16. If RNA quality is high ($A_{260}/A_{280} \sim 1.7$–2.0), then proceed to reverse transcription for RT-PCR or library prep for miRNA sequencing.

3.1.2 Reverse Transcription (RT) Reaction

1. Prepare RT reaction mix with a tube that contains the following per reaction:
   (a) 0.15 μL of 100 mM dNTPs (with dTTP).
   (b) 1.00 μL of MultiScribe™ Reverse Transcriptase, 50 U/μL.
   (c) 1.50 μL of 10X Reverse Transcription Buffer.
   (d) 0.19 μL of RNase Inhibitor, 20 U/μL.
   (e) 4.16 μL of nuclease-free water.
2. Combine 7 μL of the RT reaction mix with 5 μL of 1–10 ng of total RNA (template) and 3 μL of 5X RT Primer (TaqMan miRNA assay) in PCR tubes/plates.
3. Place the tubes/plates into a thermal cycler then incubate with the following reaction conditions:

(a) Reverse transcription:	16 °C, 30 min.
	42 °C, 30 min
(b) Heat inactivation:	85 °C, 5 min.
(c) Hold:	4 °C, hold.

3.1.3 Real-time PCR

1. Gently mix the Taqman™ Master Mix in the bottle prior to use.
2. Prepare RT-PCR master mix with enough master mix for triplicate samples with the following components (per reaction):
   (a) 1 μL of TaqMan™ miRNA assay 20× probe.
   (b) 10 μL of TaqMan™ PCR Master Mix.
   (c) 7.67 μL of nuclease-free water.
3. Load the total PCR reaction mix volume (18.67 μL) into a 96-well PCR plate and add 1.33 μL of either the cDNA template or nuclease-free water (non-template control) to each well of an appropriate PCR plate.
4. Set up the instrument to run at the following conditions:

(a) Enzyme activation:	95 °C, 10 min.
(b) Denature & Anneal/Extend (repeat for 40 cycles):	95 °C, 15 sec.
	60 °C, 1 min

5. C_T values can be subsequently analyzed to evaluate the relative expression of miRNAs.

3.2 Validating the Functional Significance of miRNAs Involved in Ferroptosis

In many diseases, miRNA expression can be dysregulated and restoring their expression to normal levels may reveal their biological impact in pathways such as ferroptosis. Candidate miRNAs can also be overexpressed or inhibited to further evaluate their phenotypic impact in cells. Lipid nanoparticle-based transfection of miRNAs or anti-miRs is a key technique employed to increase or decrease the expression of miRNAs in cells. An anti-miR is a small, chemically modified oligonucleotide that is designed to bind to and inhibit the activity of a specific miRNA [30]. Lipid nanoparticle-based transfection allows for the delivery of these small RNAs into the cell. After allowing time for change in gene activity to occur, a variety of functional assays can be performed. These include western blot to evaluate the impact on proteins involved in ferroptosis or assess the induction or inhibition of ferroptosis itself.

1. Plate cells onto a 6-well plate at appropriate density (typically around 1.0×10^6 cells/well) and incubate at 37 °C overnight.
2. Dilute oligonucleotide (miRNA or anti-miR) to desired concentration (starting range ~ 2–50 nM is typical) to a final volume of 185 μL.
3. Add desired amount of Oligofectamine™ Reagent (2–4 μL using lowest concentration that achieves the desired effect) to a separate tube and dilute to a final volume of 15 μL with Opti-MEM. Incubate for 10 min at room temperature.
4. Combine both the diluted oligonucleotide and the diluted Oligofectamine mixtures (final volume of 200 μL) and incubate for 15 min at room temperature.
5. While the samples are incubating, remove the media from cells and wash once with PBS.
6. Add 800 μL of serum-free media to each well.
7. After incubation of the mixes is complete, add the entire 200 μL mixture dropwise onto the cells.
8. Incubate at 37 °C for 4–5 h.
9. Add 1 mL of media containing 3× the normal concentration of serum to each well.
10. Incubate at 37 °C for 24–72 h and perform functional assay of choice (i.e., flow cytometric analysis, analysis of protein expression via western blot, etc.).

3.3 Evaluating miRNA Targets by Luciferase Reporter Assay

Luciferase reporter assays are a sensitive and quantitative method for measuring the interaction between a miRNA and a target mRNA. The assay uses a luciferase gene construct containing the predicted miRNA targeting sequence from the target gene, often located in the 3′UTR, a firefly luciferase reporter gene (*luc*), and typically contains selectable marker genes, such as for puromycin. The 3′UTR sequence of the candidate target gene of interest can be cloned into a backbone plasmid or the final target gene construct (s) or purchased as purified plasmid. These target site constructs are co-transfected with reporter vectors that contain a constitutive promoter (i.e., SV40, TK, or CMV), the Renilla luciferase gene (*Rluc*), followed by a late SV40 polyadenylation signal. When performing the assay, the amount of luminescence produced is proportional to the amount of miRNA that is bound to the target gene and can be used to predict the affinity of a miRNA for its target mRNA. Because miRNA:mRNA interactions can be cell context-dependent, it may be important to also consider performing the luciferase reporter assay in cell lines that model the disease of interest [31]. The described technique can be scaled to test miRNA:mRNA interaction in a host of different cell models and for a variety of different interaction pairs, with the resource-limiting step being the generation of the target gene constructs.

3.3.1 Co-transfection of Plasmids

1. Plate 1.0×10^4–2.0×10^4 cells onto a 96-well plate.
2. In separate tubes, prepare triplicate transfection conditions with 100 ng of the target gene reporter construct(s) (pGL4-GeneX, where GeneX represents your gene of interest) and 1 ng of Renillia luciferase vector (pRL-SV40). The following conditions should be used with 10 nM of miRNA or control non-specific miRNA:
   (a) pGL4-GeneX + miRNA (test condition).
   (b) pGL4-mutantGeneX + miRNA (negative control).
   (c) pGL4-GeneX + control non-specific miRNA (negative control).
   (d) pGL4-GeneX only (positive luciferase control).
3. Dilute DharmaFECT Duo (volume can range depending on the sensitivity of the cell line) in serum-free media to a final volume of 140 μL. Mix and incubate plate for 5 min at room temperature.
4. Add 35 μL of the DharmaFECT Duo mixture to the plasmid-containing tubes. Mix gently and incubate plate for 20 min at room temperature.
5. Add 280 μL of complete media to each mixture.
6. Aspirate media from the 96-well plate and add 100 μL of each condition to separate wells in triplicate.
7. Incubate at 37 °C for 24 h.

3.3.2 Dual-Luciferase Reporter Assay

1. Prepare Luciferase Assay Reagent II (LAR II) and 1× Stop & Glo® Reagent prior to experiment.
2. Dilute 5× Passive Lysis Buffer in PBS and add 20 μL of 1× solution to each well.
3. Mix in a 96 well plate that is placed on an orbital plate sharker at room temperature for 15 min.
4. Add 100 μL of LARII directly to the plate with samples.
5. Measure firefly luciferase activity.
6. Add 100 μL of Stop & Glo® Reagent to the plate with samples.
7. Measure Renilla luciferase activity.
8. For each sample, normalize firefly luciferase activity to Renilla luciferase activity and normalize the inhibition by the miRNA to the control miRNA.

4 Notes

1. Use RNaseZap to clean surfaces and pipettes prior to working with RNA. Only use molecular-grade RNase-free water, isopropanol, and ethanol for RNA work.

2. There are RNA isolation kits that may be used to isolate and clean-up RNA as desired in lieu of the isopropanol precipitation.
3. There are 1-step reverse transcription and RT-PCR protocols that employ similar principles that can be optimized and potentially used as a higher-throughput RT-PCR method of screening miRNAs.
4. While luciferase reporter assays are considered a gold standard technique to validate binding of miRNAs to mRNA, it may be beneficial to further validate the protein expression of the target mRNAs via techniques such as western blotting or Argonaute-crosslinking and immunoprecipitation (AGO-CLIP) [32].

References

1. Fire A, Xu S, Montgomery MK et al (1998) Potent and specific genetic interference by double-stranded RNA in Caenorhabditis elegans. Nature 391:806–811
2. Wightman B, Ha I, Ruvkun G (1993) Post-transcriptional regulation of the heterochronic gene lin-14 by lin-4 mediates temporal pattern formation in C. elegans. Cell 75:855–862
3. Lee RC, Feinbaum RL, Ambros V (1993) The C. elegans heterochronic gene lin-4 encodes small RNAs with antisense complementarity to lin-14. Cell 75:843–854
4. Lewis BP, Shih I-h, Jones-Rhoades MW et al (2003) Prediction of mammalian microRNA targets. Cell 115:787–798
5. Chipman LB, Pasquinelli AE (2019) miRNA targeting: growing beyond the seed. Trends Genet 35:215–222
6. Lewis BP, Burge CB, Bartel DP (2005) Conserved seed pairing, often flanked by adenosines, indicates that thousands of human genes are microRNA targets. Cell 120:15–20
7. Calin GA, Dumitru CD, Shimizu M et al (2002) Frequent deletions and down-regulation of micro-RNA genes miR15 and miR16 at 13q14 in chronic lymphocytic leukemia. Proc Natl Acad Sci 99:15524–15529
8. Calin GA, Croce CM (2006) MicroRNA signatures in human cancers. Nat Rev Cancer 6: 857–866
9. Barwari T, Joshi A, Mayr M (2016) MicroRNAs in cardiovascular disease. J Am Coll Cardiol 68:2577–2584
10. Junn E, Mouradian MM (2012) MicroRNAs in neurodegenerative diseases and their therapeutic potential. Pharmacol Ther 133:142–150
11. Xie B, Guo Y (2021) Molecular mechanism of cell ferroptosis and research progress in regulation of ferroptosis by noncoding RNAs in tumor cells. Cell Death Dis 7:101
12. Zhang X, Wang L, Li H et al (2020) Crosstalk between noncoding RNAs and ferroptosis: new dawn for overcoming cancer progression. Cell Death Dis 11:580
13. Kong Y, Chang S, Hu L et al (2018) FPN1, targeted by Mir-17-5p, regulates cellular iron homeostasis via activating Nrf2/Mir-17-5p/FPN1 axis in preclinical model of multiple myeloma. American Society of Hematology, Washington, DC
14. Luo M, Wu L, Zhang K et al (2018) miR-137 regulates ferroptosis by targeting glutamine transporter SLC1A5 in melanoma. Cell Death Differentiation 25:1457–1472
15. Tang D, Kroemer G (2020) Ferroptosis. Curr Biol 30(21):R1292–R1297
16. Xu Z, Chen L, Wang C et al (2021) MicroRNA-1287-5p promotes ferroptosis of osteosarcoma cells through inhibiting GPX4. Free Radic Res 55:1119–1129
17. Xu P, Wang Y, Deng Z et al (2022) MicroRNA-15a promotes prostate cancer cell ferroptosis by inhibiting GPX4 expression. Oncol Lett 23:1–8
18. Liu L, Yao H, Zhou X et al (2022) MiR-15a-3p regulates ferroptosis via targeting glutathione peroxidase GPX4 in colorectal cancer. Mol Carcinog 61:301–310
19. Koppula P, Zhuang L, Gan B (2021) Cystine transporter SLC7A11/xCT in cancer: ferroptosis, nutrient dependency, and cancer therapy. Protein Cell 12:599–620

20. Ni H, Qin H, Sun C et al (2021) MiR-375 reduces the stemness of gastric cancer cells through triggering ferroptosis. Stem Cell Res Ther 12:1–17
21. Rupaimoole R, Slack FJ (2017) MicroRNA therapeutics: towards a new era for the management of cancer and other diseases. Nat Rev Drug Discov 16:203–222
22. Fesler A, Liu H, Ju J (2018) Modified miR-15a has therapeutic potential for improving treatment of advanced stage colorectal cancer through inhibition of BCL2, BMI1, YAP1 and DCLK1. Oncotarget 9:2367
23. Wu N, Fesler A, Liu H et al (2018) Development of novel miR-129 mimics with enhanced efficacy to eliminate chemoresistant colon cancer stem cells. Oncotarget 9:8887
24. Berezikov E, Cuppen E, Plasterk RH (2006) Approaches to microRNA discovery. Nat Genet 38:S2–S7
25. Dave VP, Ngo TA, Pernestig A-K et al (2019) MicroRNA amplification and detection technologies: opportunities and challenges for point of care diagnostics. Lab Investig 99: 452–469
26. Pritchard CC, Cheng HH, Tewari M (2012) MicroRNA profiling: approaches and considerations. Nat Rev Genet 13:358–369
27. Linsen SE, de Wit E, Janssens G et al (2009) Limitations and possibilities of small RNA digital gene expression profiling. Nat Methods 6: 474–476
28. Agarwal V, Bell GW, Nam J-W et al (2015) Predicting effective microRNA target sites in mammalian mRNAs. elife 4:e05005
29. Enright A, John B, Gaul U et al (2003) MicroRNA targets in Drosophila. Genome Biol. BioMed Central
30. Elmen J, Lindow M, Silahtaroglu A et al (2008) Antagonism of microRNA-122 in mice by systemically administered LNA-antimiR leads to up-regulation of a large set of predicted target mRNAs in the liver. Nucleic Acids Res 36:1153–1162
31. Erhard F, Haas J, Lieber D et al (2014) Widespread context dependency of microRNA-mediated regulation. Genome Res 24:906–919
32. Chi SW, Zang JB, Mele A et al (2009) Argonaute HITS-CLIP decodes microRNA–mRNA interaction maps. Nature 460:479–486

Chapter 21

Detection of Ferroptosis in Models of Brain Diseases

Danmin Shen, Fei Yang, and Qian Li

Abstract

Ferroptosis is a regulated form of non-apoptotic cell death driven by iron-dependent lipid peroxidation. In the past decade, ferroptosis has been reported to be involved in the pathological role in the central nervous system degenerative diseases (e.g., Alzheimer's disease, Huntington's disease, and Parkinson's disease), stroke, traumatic brain injury, and brain tumor. However, how to reliably detect and classify ferroptosis from other cell death in pathological conditions remains a great challenge, especially in primary brain cells and brain tissues. Here, we summarize the methods and protocols (such as real-time PCR, western blotting, immunofluorescence staining, lipid peroxidation assay kits and probe, immunofluorescence staining, GPX activity and glutathione depletion assay kits, iron detection, and TEM) used in the present study to detect and classify ferroptosis in the brain.

Key words Ferroptosis, Central nervous system diseases, Brain tissue

1 Introduction

Ferroptosis was first coined in 2012 as a form of nonapoptotic cell death characterized by the accumulation of large amounts of iron and lipid peroxidation [1]. Over the past decade, many researchers have identified key regulators of ferroptosis, including the system Xc−/glutathione/glutathione peroxidase 4 (GPX4) axis, and several GPX4-independent systems, such as ferroptosis suppressor protein 1 (FSP1)/CoQ10, dihydroorotate dehydrogenase (DHODH), and GTP cyclohydrolase 1 (GCH1)/tetrahydrobiopterin (BH4) pathway [2, 3]. Furthermore, ferroptosis intersects with metabolism, reactive oxygen species (ROS) biology, and iron regulation, making it a unique form of cell death [4].

Through the dedicated efforts of researchers, ferroptosis has increasingly been recognized as an important process in the pathogenesis of various organs and systems, including the central nervous system (CNS). Ferroptosis is involved in several CNS diseases, such as brain tumors, traumatic brain injury, stroke, and

Guido Kroemer and Daolin Tang (eds.), *Ferroptosis: Methods and Protocols*, Methods in Molecular Biology, vol. 2712,
https://doi.org/10.1007/978-1-0716-3433-2_21,

neurodegenerative diseases (e.g., Huntington's disease, Alzheimer's disease, Parkinson's disease, and amyotrophic lateral sclerosis) [5–9].

However, detecting ferroptosis in animal and human brain tissues remains a major challenge in ferroptosis research. The cellular complexity of tissue samples, the interference of extraction and preparation processes, and the specificity of ferroptosis detection methods all limit the application of ferroptosis detection in the brain. Therefore, a combination of multiple markers (iron metabolism, lipid peroxidation, and cell metabolism) and various detection technologies are necessary to definitively identify intracellular ferroptosis. Additionally, omics techniques, particularly lipidomics, metabolomics, transcriptomics, and multi-omics techniques, are increasingly being applied in ferroptosis research and are expected to become important paradigms in future studies. In this chapter, we introduce some commonly used protocols for identifying ferroptosis in brain cells and tissues.

2 Materials

2.1 Equipment

1. Ice box.
2. Perfusion pump.
3. Tissuelyser.
4. Plate shaker.
5. Refrigeration centrifuges.
6. Spectrometer.
7. 7500 Fast Real-Time PCR system.
8. Western blot chemiluminescence imager.
9. Microtome cryostat.
10. Microscope.
11. Microscope camera.
12. Elan DRC II inductively coupled plasma mass spectrometer.
13. Transmission electron microscope.

2.2 Reagents

1. Phosphate-buffered saline (PBS), stored at 4 °C.
2. 4% paraformaldehyde, stored at 4 °C.
3. 20% and 30% sucrose solutions in PBS.
4. TBST: Tris buffered saline buffer with tween 20 (137 mM NaCl, 20 mM Tris, 0.1% Tween-20, pH 7.6).
5. Blocking solution for western blotting: 5% non-fat milk in TBST.

6. 0.01 M citric acid buffer: 9 mL 0.1 mol/L citric acid and 41 mL 0.1 mol/L sodium citrate in 450 mL H_2O (pH 6.0).
7. Primary antibody solution:
   (a) Anti-malondialdehyde antibody (ab243066, Abcam) (1: 1000) in TBST for western blotting or (1:100) in PBS with 0.3% Triton X-100 and 1% goat serum for immunofluorescence staining.
   (b) Anti-4 hydroxynonenal (4-HNE) antibody (ab48506, Abcam) (1/25 for immunofluorescence staining) in PBS with 0.3% Triton X-100 and 1% goat serum.
8. Secondary antibody solution
   (a) Species-appropriate HRP-linked secondary antibody (1: 1000–1:2000, cell signaling technology) in TBST.
   (b) Alexa Fluor 594 goat-anti-rabbit IgG/594 goat-anti-mouse IgG/488 goat-anti-rabbit IgG/Alexa Fluor 488 goat-anti-mouse IgG (1:1000–1:2000, Thermo Fisher) in PBS with 0.3% Triton X-100 and 1% goat serum.
9. BCA protein assay kit (23,221, Thermo Fisher).
10. Image-iT lipid peroxidation kit (C10445, Thermo Fisher).
11. Lipid peroxidation (MDA) assay kit (colorimetric/fluorometric) (ab118970, Abcam).
12. Lipid peroxidation (4-HNE) assay kit (ab238538, Abcam).
13. Glutathione peroxidase assay kit (ab102530, Abcam).
14. GSH/GSSG ratio detection assay kit (Fluorometric-Green) (ab138881, Abcam).

2.3 Consumables

1. 15 and 50 mL centrifuge tube.
2. 1.5 mL microtube.
3. Slide glass.
4. Glass coverslips.

3 Methods

3.1 Gene and Protein Expression Assay

The process of ferroptosis is regulated by a wide range of genes that encode proteins involved in multiple metabolic changes. For instance, during ferroptosis, the expression of genes encoding proteins such as ChaC glutathione-specific gamma-glutamylcyclotransferase 1 (CHAC1), prostaglandin-endoperoxide synthase 2 (PTGS2), solute carrier family 7 member 11 (SLC7A11), acyl-CoA synthetase long-chain family member 4 (ACSL4), glutathione peroxidase 4 (GPx4), and ferroptosis suppressor protein 1 (FSP1) are induced, while the gene expression of regulator of G protein signaling 4 (RGS4) is downregulated [7, 10–14]. Recently, antibodies against transferrin receptor 1 (TfR1) have been shown to have good specificity and sensitivity for the detection of ferroptosis in cell level and in vivo [15]. Additionally, iron metabolism-related proteins such as iron regulatory proteins (IRP1), divalent metal transporter 1 (DMT1), and ferroportin1 (Fpn1) have been shown to be altered in neurodegenerative diseases [16].

Quantitative polymerase chain reaction (qPCR) and western blotting are two commonly used methods to classify ferroptosis from other cell death. qPCR is used to detect mRNA levels, while western blotting is used to detect protein levels.

qPCR involves three basic reaction steps: denaturation, annealing, and extension. By repeating the cycle, a small amount of DNA can be rapidly amplified into millions of copies, enabling the detection of specific DNA segments. Real-time PCR techniques use fluorescence signal accumulation to monitor the whole PCR process in real-time and obtain qualitative and quantitative results for a specific nucleic acid. The results are calculated by calculating the Ct value for each sample and comparing it to a standard curve.

Western blotting, also known as protein blotting or immunoblotting, is an experimental method that detects proteins qualitatively or quantitatively based on a specific immune response between an antigen and an antibody. Western blotting combines the advantages of high-resolution gel electrophoresis with the high specificity and sensitivity of solid-phase immunoassay to identify and quantitatively detect specific antigens from complex mixtures.

3.1.1 Real-Time PCR

1. Anesthetize the mice using isoflurane and then perfused with iced PBS (pH 7.4) to remove blood and clots. Or harvest the necessary number of cells and wash the cells with cold PBS.
2. Lyse samples on ice with Trizol reagent.
3. Homogenize the tissue/cells and keep the tube on ice.
4. Transfer the homogenized tissue/cells to a cold 1.5 mL microcentrifuge tube.
5. Add 0.2 mL of chloroform, mix it upside down 10–15 times and let stand at room temperature for 10 min.
6. Centrifuge at 12,000 *g* for 15 min at 4 °C.
7. Carefully transfer the upper water phase to a new 1.5 mL microcentrifuge tube, add 0.5 mL isopropanol, mix well and leave at room temperature for 10 min.
8. Centrifuge at 12,000 *g* for 10 min at 4 °C.
9. Add 1 mL of 75% ethanol to each tube to clean the sediment.
10. Centrifuge at 7500 *g* for 5 min at 4 °C.

11. Discard the supernatant, remove the cap, and dry in a fume hood at room temperature for about 5 min.
12. Add Rnase-free water to dissolve the RNA.
13. Detect the purity and concentration of RNA with a spectrophotometer.
14. Obtain the cDNA following the manufacturer's instructions of reverse transcription kit (HiScript II Q Select RT SuperMix for qPCR, R233). The protocol is described in detail as follows:
    (a) Add 4× gDNA wiper mix (4 μL) and RNA (1 μg) into a new Rnase-free microcentrifuge tube.
    (b) Add RNase-free ddH_2O to each reaction to bring the final reaction volume to 16 μL. Reaction at 42 °C for 2 min.
    (c) Add 5× HiScript II Select qRT SuperMix II (4 μL). Reaction at 50 °C for 15 min followed by 85 °C for 5 s.
15. A modified Real-time PCR protocol from Applied Biosystems™ 7500 Fast Real-time PCR System and SYBR™ Green QPCR Mix (Thermo Fisher, 4367659) is described in detail as follows:
    (a) The experimental reaction is prepared by combining the components of the reagent mixture in the order listed in the table below. Prepare a single reagent mixture for replicating the reaction using a multiple of each component.
    (b) Gently mix the reagent mixture without creating bubbles, and then distribute the mixture to the experimental reaction tube.
    (c) Add 1 μL of experimental DNA to each reaction to bring the final reaction volume to 20 μL.
    (d) Mix the reaction without creating bubbles and then centrifuge briefly.
    (e) To set the PCR parameters, refer to the Applied Biosystems 7500 Fast Real-Time PCR System product manual.
    (f) Place the reaction in the instrument and run the program.
    (g) To obtain the CT values of the samples.
16. Use Excel to calculate the value of $2^{-\Delta\Delta Ct}$ of the sample and normalize target mRNA levels.

3.1.2 Western Blotting

1. Anesthetize the mice using isoflurane and then perfused with iced PBS (pH 7.4) to remove blood and clots. Or harvest as many cells as necessary and wash the cells with cold PBS.
2. Lysis the sample on ice with a RIPA lysis buffer containing a 10% protease inhibitor cocktail.

3. Homogenize the tissue/cells and leave the tube on ice for 30 min.
4. Transfer the homogenized tissue/cells to a cold 1.5 mL microcentrifuge tube.
5. Spin at 13,400 *g* for 5–10 min at 4 °C.
6. Transfer the supernatant to a new tube and keep on ice.
7. Quantify the protein concentration using the BCA protein assay kit (23221, Thermo Fisher, USA).
8. Add 4× loading buffer to the samples.
9. Heat samples to 95–100 °C for 5 min, then cool it on ice.
10. Load equivalent protein (20–40 μg) samples onto a polyacrylamide gel electrophoresis (SDS-PAGE) gel.
11. Load the molecular weight markers.
12. Run the gel at 80 V for about 20 min, then increase the voltage to 120–150 V for about 60 min.
13. Transfer the protein from the gel to the PVDF membrane at a constant current of 250 mA for about 90 min.
14. Incubate the membrane in blocking solution for 30 min at room temperature.
15. Incubate the membrane with appropriate diluted primary antibody and agitate gently overnight at 4 °C.
16. Wash with TBST 3 times, each time for 5 min.
17. Incubate the membrane with the species-appropriate HRP-linked secondary antibody (1:1000–1:2000) and agitate gently for 1 h at room temperature.
18. Wash 3 times with TBST for 5 min each.
19. The immunoreactive bands are visualized with a chemiluminescent substrate.
20. The chemiluminescence signal is processed using a film or digital imaging system.
21. Quantify the intensity of the bands of the target protein levels with an image analysis software.

3.2 Lipid Peroxidation

Lipid peroxidation, a hallmark of ferroptosis, refers to the oxidative degradation of lipids. Quantifying lipid peroxidation is crucial for assessing ferroptosis [17]. The BODIPY581/591 C11 reagent, which is oxidized by lipid peroxidation on the membrane, has often been used to detect lipid peroxidation in living cells of the central nervous system diseases [18, 19]. After the reagent oxidation, the peak fluorescence emission wavelength shifts from ~590 nm to ~510 nm, and the cell fluorescence changes from red to green, a change that can be collected by microscopy and flow cytometry and provides an indication of the ratio of lipid peroxidation rates.

Lipid peroxidation produces reactive aldehydes such as malondialdehyde (MDA) and 4-hydroxynonenal (4-HNE). MDA and 4-HNE are often used as markers of lipid peroxidation and indicators of ferroptosis in the brain [20–22]. MDA in the brain sample reacts with thiobarbituric acid (TBA) to generate an MDA-TBA adduct, which can be easily quantified colorimetrically (OD = 532 nm) or fluorometrically (Ex/Em = 532/553 nm). Immunohistochemical techniques use the principle of antigen-antibody binding to detect specific proteins and enable visualization of test results. It has been used to quantify MDA and 4-HNE adducts in the brain [19, 20, 23, 24]. In addition, 4-HNE can also be detected by western blotting [10]. Overall, these methods quantify the by-products of lipid peroxidation, which are crucial for assessing ferroptosis in brain cells.

3.2.1 C11-BODIPY

A modified Image-iT lipid peroxidation kit protocol (based on typical cell culture) from Thermo Fisher Scientific (https://www.thermofisher.com/order/catalog/product/C10445#/C10445) is described in detail as follows:

1. Isolate brain cells from the brains of mouse pups. Plate cells at a desired density and incubate them overnight at 37 °C.
2. Treat the cells with the compound of interest and incubate for the recommended time.
3. Add Image-iT® Lipid Peroxidation Sensor (Component A) at a final concentration of 10 mM to the cells. Then incubate for another 30 min at 37 °C.
4. Remove the media and wash the cell three times with PBS.
5. The fluorescence at individual wavelengths is read out: one at excitation/emission of 581/591 nm (Texas Red® filter set) for the reduced dye, and the other at excitation/emission of 488/510 nm (traditional FITC filter set) for the oxidized dye (*see* **Note 1**).
6. For fluorescence imaging analysis, traditional Texas Red® (590 nm) and FITC (510 nm) emission filters were used. Determine lipid peroxidation by quantitating the fluorescence intensities and calculating the ratio of intensity in Texas Red® channel to the intensity in FITC channel.
7. The ratio of the emitted fluorescence intensity at 590 nm to 510 nm gives the readout of lipid peroxidation in the cell.

3.2.2 MDA Assay

A modified lipid peroxidation (MDA) assay protocol (based on brain samples) from Abcam (https://www.abcam.cn/lipid-peroxidation-mda-assay-kit-colorimetricfluorometric-ab118970.html) is described in detail as follows:

1. Anesthetize the mice using isoflurane and then perfused with iced PBS (pH 7.4) to remove blood and clots. Or harvest the necessary number of cells (2×10^6), and wash the cells with cold PBS.
2. Rapidly homogenize brain tissue (initial recommendation: 1–10 mg) or cells in 303 μL Lysis Solution (Buffer + BHT) with a Dounce homogenizer sitting on ice, with 10–15 passes.
3. Centrifuge at 13,000 × *g* for 10 min to remove the insoluble material. Collect the supernatant (*see* **Note 2**).
4. Prepare standard curve dilutions for colorimetric or fluorometric assays.
5. Add 600 μL of TBA reagent into each vial or well containing 200 μL standard and 200 μL sample.
6. Incubate at 95 °C for 60 min. Cool to room temperature in an ice bath for 10 min.
7. Take 200 μL of the reaction mix (containing MDA-TBA adduct) and add into a 96-well microplate for analysis.
8. Immediately measure the absorbance on a microplate reader at OD 532 nm for colorimetric assay and RFU at Ex/Em = 532/553 nm for fluorometric assay.
9. Use plate reader software or Excel to construct standard curves. Calculate the trendline equation based on your standard curve data. The corrected sample OD readout is applied to the standard curve to obtain the MDA in the sample well.

3.2.3 4-HNE Assay

A modified lipid peroxidation (4-HNE) assay protocol (based on brain samples) from Abcam (https://www.abcam.com/lipid-peroxidation-4-hne-assay-kit-ab238538.html) is described in detail as follows:

1. Anesthetize the mice using isoflurane and then perfused with iced PBS (pH 7.4) to remove blood and clots. Or harvest the necessary number of cells (2×10^6), and wash the cells with cold PBS.
2. Homogenize the tissue/cells with iced PBS and keep the tube on ice.
3. Centrifuge at 5000 × *g* for 5 min at 4 °C.
4. Transfer the supernatant to a new tube and keep on ice.
5. Dilute samples in 1× PBS containing 0.1% BSA before adding. Incubate at room temperature for 10 min on an orbital shake.
6. Add 50 μL of sample or 4-HNE-BSA standard to the wells of the 4-HNE Conjugate coated plate.
7. Add 50 μL of the diluted anti-4-HNE antibody to each well, incubate at room temperature for 1 h on an orbital shaker.

8. Wash 3 times with 250 μL of 1× Wash Buffer with thorough aspiration between each wash. After the last wash, empty wells and tap microwell strips on absorbent pad or paper towel to remove excess 1× Wash Buffer.
9. Add 100 μL of the diluted Secondary Antibody-HRP Conjugate to all wells and incubate for 1 h at room temperature on an orbital shaker.
10. Wash the strip 3 times according to **step 8** above.
11. Warm the Substrate Solution to room temperature. Add 100 μL of Substrate Solution to each well. Incubate at room temperature for 2–20 min on an orbital shaker (*see* **Note 3**).
12. Stop the enzyme reaction by adding 100 μL of Stop Solution to each well.
13. Read out the absorbance of each well immediately on a microplate reader with 450 nm as the primary wavelength.
14. Construct standard curves using plate reader software or Excel. Calculate the trendline equation based on your standard curve data. The corrected sample OD readout is applied to the standard curve to obtain the 4-HNE adduct concentration in the sample trap.

3.2.4 Immunofluorescence Staining for Detection of MDA and/or 4-HNE

1. For the cell samples, 4% paraformaldehyde is applied for 10 min and the cells are washed with PBS three times.
2. Anesthetize the mice using isoflurane and cardiac perfused with 0.01 M PBS (4 °C), followed by 4% paraformaldehyde.
3. Postfix the mouse brains in 4% paraformaldehyde overnight at 4 °C and then dehydrate in 20% and 30% sucrose solutions for 72 h at 4 °C successively. Embed the brain tissues with OCT and store at −80 °C.
4. Cut the frozen tissues into consecutive coronal sections (16 μm).
5. Bring sections of slides from −80 °C to room temperature. Wait 30 min until the slide is dry.
6. Wash the slides with 0.01 M PBS three times for 5 min then develop an antigen-retrieval protocol using a high temperature 0.01 M citric acid buffer (pH 6.0) for 10 min.
7. Incubate the sections in 1% Triton X-100 in PBS at room temperature for 15 min, followed by three PBS washes.
8. Incubate the sections in 0.3% H_2O_2 for 10 min followed by PBS.
9. Block the slides/cells at room temperature for 1 h with 5% goat serum in 0.3% Triton X-100. Incubate the sections at 4 °C overnight with primary antibodies for MDA and/or 4-HNE.

10. Wash the sections with PBS and incubate the sections at room temperature for 1 h with fluorescence-conjugated secondary antibodies.
11. After washing with PBS, mount the sections in the VECTA-SHIELD mounting medium with DAPI solution.
12. Acquired images with a microscope and quantify the mean fluorescence intensity in a double-blind manner using the Image J software (*see* **Note 4**).

3.3 GPX Activity Assay

GPX4 is a selenoprotein that plays a critical role in counteracting lipid oxidation in the membrane via its GSH-dependent peroxidase activity [25]. GPX4 has long been considered a core protein in the evaluation of ferroptosis in the brain [26, 27]. The Glutathione Peroxidase Assay Kit is commonly used to quantify the activity of all GSH-dependent peroxidases in a sample. This assay involves the oxidation of GSH by GPx to produce oxidized glutathione (GSSG) during the reduction of cumene hydroperoxide. The subsequent reduction of GSSG to GSH by glutathione reductase (GR) consumes nicotinamide adenine dinucleotide phosphate (NADPH), and the decrease in NADPH (measured at OD = 340 nm) is proportional to GPx activity. Additionally, the GPX4 inhibitor ML-162 can be included as a positive control to identify novel inhibitors of human GPX4 (GPX4 Inhibitor Screening Assay Kit, Cayman Item No. 701880). Some studies have utilized a specific phospholipid substrate, phosphatidylcholine hydroperoxide (PCOOH), to determine GPx4 activity in cells, but this approach has not been widely applied to tissues [28].

Total GPX activity was measured with Glutathione Peroxidase Assay Kit (ab102530, Abcam). A modified GPx4 activity assay protocol (based on brain samples) from Abcam (https://www.abcam.cn/glutathione-peroxidase-assay-kit-colorimetric-ab102530.html) is described in detail as follows:

1. Prepare colorimetric assay standard curve dilution and add 100 μL standard dilutions in standard wells.
2. Mice were anesthetized using isoflurane and then perfused with iced PBS (pH 7.4) to remove blood and clots. Or harvest the necessary number of cells (2×10^6), and wash the cells with cold PBS.
3. Resuspend samples in 200 μL cold Assay Buffer (initial recommendation: 100 mg).
4. Homogenize brain tissue or cells on ice.
5. Centrifuge for 15 min at 4 °C at 10,000 *g* using a cold microcentrifuge to remove any insoluble material.
6. Collect the supernatant and transfer it to a clean tube and keep on ice.

7. Add 2–50 μL samples into a 96-well microplate for each well.
8. Add 5–10 μL of the GPx Positive Control (adjust volume to 50 μL/well with Assay Buffer) for Positive Control wells.
9. Add 50 μL Assay Buffer for Reagent Control well.
10. Prepare and mix enough reaction reagents for the number of to be performed (40 μL reaction mix = 33 μL Assay Buffer, 3 μL 40 mM NADPH solution, 2 μL GR solution, 2 μL GSH solution).
11. Add 40 μL of Reaction Mix to sample, positive control(s) and reagent control wells.
12. Mix well and incubate at room temperature for 15 min to deplete all GSSG in the sample.
13. Measure the OD 340 nm before adding cumene hydroperoxide (If the OD at 340 nm is lower than 1.0 add more NADPH to ensure there is enough NADPH in the reaction system. 1 μL of 40 mM NADPH will give ~0.5 OD at 340 nm).
14. Add 10 μL cumene hydroperoxide solution, to the sample, positive control and reagent control wells only, to start the GPx reaction. Mix well.
15. Measure output (A1) on a microplate reader at OD340 nm at T1.
16. Incubate at 25 °C for 5 min (or longer if the GPx activity is low). Protect from light.
17. Measure output (A2) on a microplate reader at OD340 nm once every 5 min for 40 min (T2).
18. Construct standard curves using plate reader software or Excel. Calculate the trendline equation based on your standard curve. The corrected sample OD reading is applied to the standard curve to obtain the GPx activity in the sample well (*see* **Note 5**).

3.4 Quantification of GSH Content and GSH/GSSG Ratio

GSH is the smallest intracellular protein thiol molecule in the cell and plays an important role in preventing cell damage caused by reactive oxygen species such as free radicals and peroxides [20]. It is also crucial for the biogenesis of iron-sulfur clusters and is a cofactor for several enzymes, including GPX4 [29]. The depletion of GSH can not only inactivate GPX4 but also mobilize Fe(II) for Fenton chemistry, promoting the propagation of lipid peroxides and ultimately leading to ferroptosis [30]. Glutathione exists in reduced (GSH) and oxidized (GSSG) states. In healthy cells, over 90% of the total glutathione pool is in the reduced form (GSH). When cells are exposed to increased levels of oxidative stress, GSSG accumulates and the ratio of GSSG to GSH increases, indicating oxidative stress. The GSH/GSSG assay kit uses a proprietary non-fluorescent dye that becomes strongly fluorescent upon reaction with GSH. The ratio of GSSG to GSH can be calculated by reducing GSSG to GSH via GR.

A modified GSH content and GSH/GSSH ratio detection assay protocol (based on brain samples) from Abcam (https://www.abcam.cn/gshgssg-ratio-detection-assay-kit-fluorometric-green-ab138881.html) is described in detail as follows:

1. Anesthetize the mice using isoflurane and then perfused with iced PBS (pH 7.4) to remove blood and clots. Or harvest the necessary number of cells (1×10^7–10^8), and wash the cells with cold PBS.
2. Resuspend tissue (20 mg) in 400 μL of ice-cold Mammalian Lysis Buffer (alternatively you can use PBS/0.5% NP-40–0.5% NP40 can be made up in PBS pH 6.0 for lysis). Or resuspend cells in 100 μL of ice-cold Mammalian Lysis Buffer.
3. Homogenize the tissue with a Dounce homogenizer with 10–15 passes. Or quickly homogenize the cells by pipetting up and down several times.
4. Centrifuge sample for 15 min at 4 °C at top speed using a cold microcentrifuge to remove any insoluble material.
5. Collect the supernatant and transfer it to a clean tube. The sample should be kept on ice.
6. Remove enzymes from the sample by TCA/$NaHCO_3$ deproteinization:
   (a) Add 1 volume ice-cold 100% (w/v) TCA into 5 volumes of sample and vortex briefly to mix.
   (b) To incubate on ice for 5–10 min.
   (c) Centrifuge samples at 12,000 *g* for 5 min at 4 °C in a cold centrifuge and transfer supernatant to a fresh tube. Measure the volume of the supernatant.
   (d) Neutralize the sample by adding $NaHCO_3$ to the supernatant and briefly vortexing the supernatant. Add $NaHCO_3$ drop by drop until pH equals 4–6 (use pH paper to test 1 μL of the sample).
7. Bicinchoninic acid (BCA) analysis to quantify protein concentration in supernatant.
8. Prepare the standard curve dilutions for GSH and GSSH.
9. Prepare GSH assay mixture by diluting a 50× Thiol Green Stock solution in Assay Buffer at 1:50 dilution and mixing well by vortexing.
10. Prepare TGAM by diluting the 25× GSSG Probe Stock solution (Step 1.6) in GAM at 1:25 dilution and mix well by vortexing.

11. For GSH detection, add 50 μL of GSH Assay Mixture into each GSH standard and sample well to make the total assay volume 100 μL/well.
12. For Total GSH + GSSG (reduced and oxidized), add 50 μL of Total Glutathione Assay Mixture into each GSSG standard and sample well to make total assay volume 100 μL/well.
13. Incubate at room temperature for 10–60 min, protected from light.
14. A fluorescence microplate reader was used to monitor the fluorescence at Ex/Em = 490/520 nm.
15. Use a plate reader software or Excel to construct the standard curves. Calculate the trendline equation based on your standard curve data. Apply the corrected sample OD reading to the standard curve to get GSH and e total glutathione (GSH + GSSG) concentration in the sample wells. The GSSG level and the GSH/GSSG ratio can be calculated by subtracting GSH from the total glutathione level.

3.5 Iron Concentration

Iron is an essential component in driving intracellular lipid peroxidation and ferroptosis [31]. The presence of markers of iron-dependent lipid peroxidation can be used to identify the presence of ferroptosis in human and animal brain samples [32]. Perl's staining is one of the common histochemical methods for detecting non-heme iron in cells and tissues. Positive signals in cells and tissues can be continuously amplified by the addition of DAB-enhanced solutions, thus enabling the detection of the iron content and distribution in the brain [23, 29, 33, 34]. Inductively coupled plasma-mass spectrometry (ICP-MS) combines the high-temperature ionization characteristics of inductively coupled plasma with the advantages of sensitive and fast scanning of mass spectrometer, thus realizing the high sensitivity analysis of inorganic elements, including iron. Due to its high sensitivity to intracerebral iron, ICP-MS can be used to monitor iron in brain tissue during ferroptosis [18].

The oxidation state of iron, on the other hand, is important for its ability to contribute to ferroptosis. Ferrous iron (Fe^{2+}) promotes ferroptosis, ferric iron (Fe^{3+}) is generally stored in ferritin [35]. Thus, determining the redox state of iron is as important as detecting iron overload for ferroptosis [20, 36]. The ferric carrier protein dissociates ferric acid into a solution in the presence of an acid assay buffer. Free ferrous iron reacts with Iron Probe to produce a stable-colored complex with absorbance at 593 nm. Ferric iron can be reduced to form Fe^{2+} enabling the measurement of total iron (Fe^{2+} and Fe^{3+}). The ferric iron level is calculated by subtracting the ferrous iron from the total iron [23, 33]. Therefore, the application of an iron colorimetric assay kit is useful to sense the redox state of iron in brain tissue and thus to detect ferroptosis. In addition, FerroOrange, a novel probe capable of tracking intracellular Fe^{2+} in living cells, has been used in the study of central nervous system diseases in cells [12].

3.5.1 Perl's Staining

1. Anesthetize the mice using isoflurane and cardiac perfuse the mice with 0.01 M PBS (4 °C), followed by 4% paraformaldehyde.
2. Postfix the brains in 4% paraformaldehyde overnight at 4 °C and then dehydrate the brains in 20% and 30% sucrose solutions for 72 h at 4 °C successively. Embed the brain tissues with OCT and store at −80 °C.
3. Cut the frozen tissues into consecutive coronal sections (16 μm).
4. Bring sections of slides from −80 °C to room temperature. Wait 30 min until the slides are dry.
5. Wash the slides four times in PBS for 5 min each.
6. The brain segment was incubated in a solution containing 20% HCl and 10% Potassium Ferrocyanide for 30 min with a gentle shake.
7. Wash the slides four times in PBS for 5 min each.
8. Incubate brain sections in 3,3′-diaminobenzidine (DAB) with the DAB staining kit according to the manufacturer's protocol.
9. Wash the slides in ddH_2O for 5 min.
10. Incubate brain sections in a hematoxylin stain solution for counter-staining about 15–30 min.
11. Examine the stain under a microscope.
12. Rinse the slide five times in distilled water.
13. Dehydrate in different concentrations of ethanol and xylenes (70% ethanol 5 min, 80% ethanol 5 min, 90% ethanol 5 min, 100% ethanol 5 min for 2 times, xylenes 5 min for 2 times).
14. Mount the slides with neutral gum.
15. Acquire images by a microscope and count positive stained cells in a double-blind manner using the Image J software (*see* **Note 6**).

3.5.2 Iron Elements Determination with Inductive Coupled Plasma-Mass Spectrometry (ICP-MS)

1. Anesthetize the mice using isoflurane and then perfuse the mice with iced PBS (pH 7.4) to remove blood and clots.
2. Weigh the tissue and place it in a quartz tube.
3. Add 0.8 mL of nitric acid to each sample.
4. Pre-digestion at room temperature for 2 h.

5. Place the quartz tubes in a microwave digestion system (Ultra WAVE, Milestone, Italy).
6. Set the temperature as follows: room temperature – 150 °C (5 min), 150–190 °C (5 min), 190–190 °C (20 min).
7. Dilute obtained solutions to 8 mL with ultra-pure water (18 MΩ cm).
8. Add Indium prior to analysis as an internal standard.
9. Set instrumental parameters as follows: a nebulizer gas flow rate of 1.03 L/min, an auxiliary gas flow rate of 1.86 L/min, a plasma gas flow rate of 18.0 L/min, an RF power of 1150 w, a dwell time of 50–100 ms, a single point peak hopping, and a mass resolution of 0.7–0.9 amu.
10. Analyze iron levels of samples by ICP-MS (ELAN DRC II, PerkinElmer, USA) instrumental.

3.5.3 Iron Colorimetric Assay (Fe^{2+}, Fe^{3+}, Total Iron)

A modified iron colorimetric assay protocol (based on brain samples) from Abcam (https://www.abcam.cn/iron-assay-kit-colorimetric-ab83366.html) is described in detail as follows:

1. Anesthetize the mice using isoflurane and then perfuse the mice with iced PBS (pH 7.4) to remove blood and clots.
2. Harvest the amount of tissue necessary for each assay (initial recommendation = 10 mg tissue). Homogenize the tissue in 4–10 volumes of Iron Assay Buffer using a Dounce homogenizer sitting on ice with 10–15 passes.
3. Centrifuge at 16,000 × *g* for 10 min to remove the insoluble material.
4. Collect the supernatant and transfer it to a clean tube. Keep on ice.
5. Prepare standard curve dilutions for colorimetric or fluorometric assays.
6. Set up Reaction wells (see table below):
   - Standard wells = 100 μL Standard dilutions.
   - Sample wells = 2–50 μL samples (adjust volume to 100 μL/well with Iron Assay Buffer).
7. Add 5 μL Iron Reducer to each Standard well.
8. For iron (II) assay: add 5 μL of Assay Buffer to each sample.
9. For total iron (II + III) assay: add 5 μL of Iron Reducer to each sample.
10. Mix and incubate standards and samples at 37 °C for 30 min.
11. Add 100 μL Iron Probe to each well containing the Iron Standard and test samples.
12. Mix and incubate at 37 °C for 60 min protected from light.

13. Measure output immediately on a colorimetric microplate reader (OD 593 nm).
14. Construct standard curves using plate reader software or Excel. Calculate the trendline equation based on your standard curve data. Apply the corrected sample OD reading to the standard curve to get the concentration of Iron (Iron (II), Iron (III), and Total Iron) in the sample wells.

3.6 Transmission Electron Microscopy (TEM)

Ferroptosis exhibits distinct morphological features that distinguish it from other types of cell death. Specifically, it is characterized by mitochondrial shrinkage with increased mitochondrial membrane density and decreased mitochondrial cristae [1, 35]. Additionally, some studies have suggested that an electron-lucent nucleus may be a hallmark of ferroptosis [37]. TEM is a useful technique for observing mitochondrial morphology in brain tissue and identifying cell types. By using TEM, we can observe whether the morphological changes are classical signs of ferroptosis and provide tentative identification of cell types [29, 34, 38, 39].

1. Anesthetize the mice using isoflurane and then perfuse the mice with iced PBS (pH 7.4), followed by 4% paraformaldehyde.
2. Place the brain tissues into the culture dishes containing 2.5% pre-cooled glutaraldehyde.
3. Cut the target tissue into 1 mm cubes under stereo microscopy (*see* **Note 7**).
4. For cell samples, replace the entire culture medium before harvest, gently scrape off the cells with a cell scraper or digest it with trypsin, and quickly centrifuge the cells at 300 *g* for 10 min, and discard the supernatant.
5. Fix tissues/cells with 2.5% glutaraldehyde (pH 7.4) for 2 h.
6. Wash 3 times for 10 min each with 15 mL of 0.1 PB buffer solution.
7. Put the tissue in 1% osmic acid (pH 7.4) for 2 h.
8. Wash 3 times for 10 min each with 15 mL of 0.1 PB buffer solution.
9. Immerse the tissue in a 50, 60, 70, 80, and 90% ethanol solution for 10 min at a time.
10. Immerse in 100% ethanol solution 2 times, each time for 10 min.
11. Immerse in acetone 2 times, each time for 15 min.
12. Immerse the tissue in a 1:1 dilution of epoxy propane and epoxy resin embedding agent for 4 h.

13. Immerse the tissue in epoxy overnight.
14. Place the tissue in a 45 °C incubator for 12 h and a 60 °C incubator for 24 h.
15. Cut the tissue into sections with thickness of 50–70 nm thickness using an ultrathin microtome.
16. Stain with uranium acetate for 30 min.
17. Wash 3 times with ddH_2O.
18. Stain with lead citrate for 2 min.
19. Wash 1 time with NaOH.
20. Wash 3 times with ddH_2O.
21. Drying naturally.
22. Acquire images using an electron microscopy and count positive stained cells in a double-blind manner using the Image J software.

4 Notes

1. When using flow cytometry to detect fluorescence changes, if only red fluorescence changes and green fluorescence does not, it is recommended to shorten the sample loading time and strictly avoid light exposure.
2. For brain tissue samples, protein can be precipitated by homogenizing 1–10 mg of tissue in 150 μL ddH_2O + 3 μL BHT, followed by adding 1 volume of 2 N perchloric acid, vortexing, and centrifugation to remove precipitated protein. Take 200 μL of the supernatant from each sample and add it to a microcentrifuge tube. The protein concentration can be quantified using the BCA protein assay kit, and the concentration of MDA can be expressed in nmol/mg or nmol/μg of protein.
3. Monitor the plate carefully during the assay; if the color changes rapidly, the reaction may need to be stopped sooner to prevent saturation.
4. If low or no signal is obtained from immunofluorescence staining, confirm that the primary antibody is applicable to frozen sections and immunofluorescence staining. The primary antibody incubation time can be extended to 24 h or antigen retrieval time to 15 min. We recommend the use of confocal fluorescence microscopy to observe the staining results. If non-specific signals are observed, try using the recommended dilution on the product webpage or datasheet as a starting point for optimization. Alternatively, reduce the dilution concentration of the secondary antibody.

5. If the A1 reading is too low (<0.7), it may indicate that there is too much GPx or GSSG in the sample. Dilute the sample or remove GSSG from the sample, such as by dialysis or using a spin filter. Read A1 and A2 in the linear range of the reaction to obtain more accurate results.
6. All reagents should be prepared and used immediately.
7. Organize the material collection in a few minutes. If the tissue needs to be positioned, take it as a cuboid, but less than 1 mm^3. Keep all containers, instruments, and fixed liquids pre-cooled when working on ice.

References

1. Dixon SJ, Lemberg KM, Lamprecht MR et al (2012) Ferroptosis: an iron-dependent form of nonapoptotic cell death. Cell 149:1060–1072
2. Liang D, Minikes AM, Jiang X (2022) Ferroptosis at the intersection of lipid metabolism and cellular signaling. Mol Cell 82:2215–2227
3. Chen X, Li J, Kang R et al (2021) Ferroptosis: machinery and regulation. Autophagy 17: 2054–2081
4. Stockwell BR (2022) Ferroptosis turns 10: emerging mechanisms, physiological functions, and therapeutic applications. Cell 185:2401–2421
5. Han L, Zhou J, Li L et al (2022) SLC1A5 enhances malignant phenotypes through modulating ferroptosis status and immune microenvironment in glioma. Cell Death Dis 13: 1071
6. Liu T, Zhu C, Chen X et al (2022) Ferroptosis, as the most enriched programmed cell death process in glioma, induces immunosuppression and immunotherapy resistance. Neuro-Oncology 24:1113–1125
7. Li Q, Han X, Lan X et al (2017) Inhibition of neuronal ferroptosis protects hemorrhagic brain. JCI Insight 2:e90777
8. Stockwell BR, Friedmann Angeli JP, Bayir H et al (2017) Ferroptosis: a regulated cell death nexus linking metabolism, redox biology, and disease. Cell 171:273–285
9. Tang D, Chen X, Kang R et al (2021) Ferroptosis: molecular mechanisms and health implications. Cell Res 31:107–125
10. Ates G, Goldberg J, Currais A et al (2020) CMS121, a fatty acid synthase inhibitor, protects against excess lipid peroxidation and inflammation and alleviates cognitive loss in a transgenic mouse model of Alzheimer's disease. Redox Biol 36:101648
11. Ratan RR (2020) The chemical biology of ferroptosis in the central nervous system. Cell Chem Biol 27:479–498
12. Tuo QZ, Liu Y, Xiang Z et al (2022) Thrombin induces ACSL4-dependent ferroptosis during cerebral ischemia/reperfusion. Signal Transduct Target Ther 7:59
13. Bersuker K, Hendricks JM, Li Z et al (2019) The CoQ oxidoreductase FSP1 acts parallel to GPX4 to inhibit ferroptosis. Nature 575:688–692
14. Jian X, Zhao G, Chen H et al (2022) Revealing a novel contributing landscape of ferroptosis-related genes in Parkinson's disease. Comput Struct Biotechnol J 20:5218–5225
15. Feng H, Schorpp K, Jin J et al (2020) Transferrin receptor is a specific ferroptosis marker. Cell Rep 30:3411–23e7
16. Rouault TA (2013) Iron metabolism in the CNS: implications for neurodegenerative diseases. Nat Rev Neurosci 14:551–564
17. Bao Z, Hua L, Ye Y et al (2021) MEF2C silencing downregulates NF2 and E-cadherin and enhances Erastin-induced ferroptosis in meningioma. Neuro-Oncology 23:2014–2027
18. Shen D, Wu W, Liu J et al (2022) Ferroptosis in oligodendrocyte progenitor cells mediates white matter injury after hemorrhagic stroke. Cell Death Dis 13:259
19. Xiao Z, Shen D, Lan T et al (2022) Reduction of lactoferrin aggravates neuronal ferroptosis after intracerebral hemorrhagic stroke in hyperglycemic mice. Redox Biol 50:102256
20. Park MW, Cha HW, Kim J et al (2021) NOX4 promotes ferroptosis of astrocytes by oxidative stress-induced lipid peroxidation via the impairment of mitochondrial metabolism in Alzheimer's diseases. Redox Biol 41:101947
21. Zhu K, Zhu X, Liu S et al (2022) Glycyrrhizin attenuates hypoxic-ischemic brain damage by

inhibiting Ferroptosis and Neuroinflammation in neonatal rats via the HMGB1/GPX4 pathway. Oxidative Med Cell Longev 2022: 8438528
22. Zhu K, Zhu X, Sun S et al (2021) Inhibition of TLR4 prevents hippocampal hypoxic-ischemic injury by regulating ferroptosis in neonatal rats. Exp Neurol 345:113828
23. Rui T, Wang H, Li Q et al (2021) Deletion of ferritin H in neurons counteracts the protective effect of melatonin against traumatic brain injury-induced ferroptosis. J Pineal Res 70: e12704
24. Chen KN, Guan QW, Yin XX et al (2022) Ferrostatin-1 obviates seizures and associated cognitive deficits in ferric chloride-induced posttraumatic epilepsy via suppressing ferroptosis. Free Radic Biol Med 179:109–118
25. Ursini F, Maiorino M, Valente M et al (1982) Purification from pig liver of a protein which protects liposomes and biomembranes from peroxidative degradation and exhibits glutathione peroxidase activity on phosphatidylcholine hydroperoxides. Biochim Biophys Acta 710: 197–211
26. Seiler A, Schneider M, Forster H et al (2008) Glutathione peroxidase 4 senses and translates oxidative stress into 12/15-lipoxygenase dependent- and AIF-mediated cell death. Cell Metab 8:237–248
27. Banjac A, Perisic T, Sato H et al (2008) The cystine/cysteine cycle: a redox cycle regulating susceptibility versus resistance to cell death. Oncogene 27:1618–1628
28. Stolwijk JM, Falls-Hubert KC, Searby CC et al (2020) Simultaneous detection of the enzyme activities of GPx1 and GPx4 guide optimization of selenium in cell biological experiments. Redox Biol 32:101518
29. Bao WD, Pang P, Zhou XT et al (2021) Loss of ferroportin induces memory impairment by promoting ferroptosis in Alzheimer's disease. Cell Death Differ 28:1548–1562
30. Patel SJ, Frey AG, Palenchar DJ et al (2019) A PCBP1-BolA2 chaperone complex delivers iron for cytosolic [2Fe-2S] cluster assembly. Nat Chem Biol 15:872–881
31. Chen X, Yu C, Kang R et al (2020) Iron metabolism in Ferroptosis. Front Cell Dev Biol 8: 590226
32. Hadian K, Stockwell BR (2021) A roadmap to creating ferroptosis-based medicines. Nat Chem Biol 17:1113–1116
33. Zhang H, Ostrowski R, Jiang D et al (2021) Hepcidin promoted Ferroptosis through iron metabolism which is associated with DMT1 signaling activation in early brain injury following subarachnoid hemorrhage. Oxidative Med Cell Longev 2021:9800794
34. Xie BS, Wang YQ, Lin Y et al (2019) Inhibition of ferroptosis attenuates tissue damage and improves long-term outcomes after traumatic brain injury in mice. CNS Neurosci Ther 25: 465–475
35. Stockwell BR, Jiang X, Gu W (2020) Emerging mechanisms and disease relevance of Ferroptosis. Trends Cell Biol 30:478–490
36. Imai T, Iwata S, Hirayama T et al (2019) Intracellular Fe(2+) accumulation in endothelial cells and pericytes induces blood-brain barrier dysfunction in secondary brain injury after brain hemorrhage. Sci Rep 9:6228
37. Miyake S, Murai S, Kakuta S et al (2020) Identification of the hallmarks of necroptosis and ferroptosis by transmission electron microscopy. Biochem Biophys Res Commun 527: 839–844
38. Qin X, Zhang J, Wang B et al (2021) Ferritinophagy is involved in the zinc oxide nanoparticles-induced ferroptosis of vascular endothelial cells. Autophagy 17:4266–4285
39. Zille M, Karuppagounder SS, Chen Y et al (2017) Neuronal death after hemorrhagic stroke in vitro and in vivo shares features of ferroptosis and necroptosis. Stroke 48:1033–1043

Index

Guido Kroemer and Daolin Tang (eds.), *Ferroptosis: Methods and Protocols*, Methods in Molecular Biology, vol. 2712, https://doi.org/10.1007/978-1-0716-3433-2,

MIX
Papier aus verantwortungsvollen Quellen
Paper from responsible sources
FSC® C105338

If you have any concerns about our products,
you can contact us on
ProductSafety@springernature.com

In case Publisher is established outside the EU,
the EU authorized representative is:
Springer Nature Customer Service Center GmbH
Europaplatz 3, 69115 Heidelberg, Germany

Printed by Libri Plureos GmbH
in Hamburg, Germany